Energy, Climate Change & Sustainability in the Global South

Editor-in-Chief

Xiaolong Zou, School of International and Public Affairs, Jilin University, Changchun, Jilin, China

The "Global South" emerges as an undeniable force in global energy and climate governance, making substantial contributions to international climate negotiations and sustainable development. Representative of these emerging nations in the Global South, the BRICS+ Countries have fortified their collaboration on sustainable development through the BRICS cooperation mechanism. Actively engaging in the formulation of global energy and climate governance rules, they play a pivotal role in the sustainable development of the Global South and the entire world.

This book series centers on the BRICS+ Countries, delving into their current situations, challenges, and prospects in energy, climate, and sustainable development. It amalgamates the research findings of various experts and scholars to offer profound insights and recommendations on achieving sustainable development in the Global South.

Sustainability stands as the linchpin for achieving economic and social development in Global South countries. This book series selects the BRICS+ Countries as representative research subjects, providing comprehensive insights and suggestions on progressing in energy, climate, and sustainable development for Global South nations. It aims to serve as a theoretical reference and source of experience for advancing the development of the BRICS+ Countries and the Global South, thereby realizing global sustainable development goals.

Xiaolong Zou

Thirty Years of Global Climate Governance

From Rio to Glasgow

Springer

Xiaolong Zou
School of International and Public Affairs
Jilin University
Changchun, Jilin, China

ISSN 3005-1568 ISSN 3005-1576 (electronic)
Energy, Climate Change & Sustainability in the Global South
ISBN 978-981-95-8456-7 ISBN 978-981-95-8457-4 (eBook)
https://doi.org/10.1007/978-981-95-8457-4

This work was supported by Ministry of Education of the People's Republic of China (20YJCZH255), Jilin University (International Cooperation Platform Project "NO. TS2023017") and Social Sciences Office of Jilin Province (SKZ2023085).

This Springer imprint is published by the registered company Springer Nature Singapore Pte Ltd.
The registered company address is: 152 Beach Road, #21-01/04 Gateway East, Singapore 189721, Singapore

For my daughter
Zou Yun-an

Preface

This book is more than just a culmination of years of research and academic inquiry—it is the fruit of an intellectual odyssey that spans nearly a decade of my life. It began the moment I embarked on my journey as an Assistant Professor at the School of International and Public Affairs, Jilin University. What began as a simple academic curiosity—an academic exploration of climate change and global governance—has transformed into something much greater. It is the manifestation of countless hours spent poring over climate data, wrestling with global diplomatic shifts, and, above all, grappling with the deep moral questions that underlie the climate crisis.

I vividly remember my first encounter with the terms "climate change," "climate governance," and "COP" as a fresh graduate student. At that time, these words were abstract to me, mere terms I could not fully grasp. But as the years passed, they became woven into the fabric of my thinking, no longer just academic jargon but urgent realities that shape our world. Today, not only do I teach these concepts to my students but I also challenge them to question the very systems that govern how we respond to this unprecedented global challenge. This book is not just the work of a scholar; it is the work of a citizen of the world, someone who feels the weight of what is at stake.

This book was born in the classroom, where I was preparing for my undergraduate course on *Global Environmental and Climate Governance.* As I delved deeper into the literature, I stumbled upon a Chinese book titled 《从哥本哈根到巴黎》 (*From Copenhagen to Paris*), a work that profoundly influenced my perspective on the political intricacies of climate negotiations. The more I studied, the clearer it became: the need for a global reckoning on climate change is no longer theoretical—it is existential.

This book seeks to illuminate the journey of global climate governance, tracing its evolution from the Earth Summit in 1992 to the critical crossroads of COP26 in 2021. It examines the pivotal roles of key players—states, international organizations, non-governmental actors, and cities—in shaping climate action. Each chapter serves as a lens to understand the political, economic, and social forces that have propelled and hindered progress. From the nuanced role of the United Nations in facilitating global cooperation, to the European Union's leadership in setting ambitious climate targets, from China's transformation from a reluctant participant to a climate powerhouse, to the dynamic coalitions of states that have emerged over the years, this book paints a vivid portrait of a global movement in flux.

The chapters explore pivotal national and regional contributions, dissecting the role of the United States, the United Kingdom, India, ASEAN countries, and the evolving influence of NGOs like WWF. I delve into the delicate balance of ambition and realism that has defined international negotiations and the tensions between climate equity and economic development. At the heart of this exploration is the intersection of policy and power—where geopolitical tensions, historical inequities, and the urgent call for climate justice collide.

But this book is not just a narrative of state-centric negotiations. It also gives voice to the non-state actors—the cities, the businesses, the civil society organizations—that are increasingly shaping the climate discourse. In the age of "bottom-up" governance, cities like Kitakyushu in Japan, and transnational networks like C40, are proving that local action can drive global change. The role of non-governmental organizations, particularly in amplifying marginalized voices and pushing the global agenda forward, is also critically examined. Through their activism, advocacy, and practical innovations, NGOs have become the engines of change, not just advocating for policy shifts but also implementing real-world solutions at the grassroots level.

This book also confronts the persistent contradictions within global climate governance. Despite the clear, unequivocal scientific consensus on climate change, the political will to act remains woefully inadequate. We face a structural divide between the urgent need for transformative action and the slow grind of diplomatic processes, where real-world consequences clash with political compromises. Yet, amidst this struggle, there are triumphs—significant milestones like the Paris Agreement and growing momentum for carbon pricing and green finance. These moments remind us that while the road ahead is fraught with challenges, there are paths forward, and there are victories to be won.

In delving deep into these themes, I hope to inspire a new generation of scholars, activists, and policymakers who understand that climate governance is not just a theoretical exercise. It is a moral, social, and political battle for our future. My wish is that this book serves as a call to action—an urgent plea to all who care about the future of the planet to not only understand the complexity of global climate governance but to actively shape its future.

As we stand at the crossroads of this global challenge, this book is a testament to the passionate, persistent effort required to overcome the political inertia that has too long defined the climate debate. But it is also a hopeful reminder: change is possible. We have made strides, but the real work—the hardest work—is yet to come. And it is work we must all embrace, for the sake of our children, our planet, and the generations to come.

Changchun, China
November 2025

Xiaolong Zou

Acknowledgements Before I dive into the academic and professional acknowledgements, I must start with the deepest and most heartfelt thanks to the people who stood by me throughout this monumental task. Without their unwavering support, this book would have remained nothing more than a distant dream—perhaps relegated to the depths of my *to-write* list, right next to "learn German" and "get more sleep."

First and foremost, I must express my profound gratitude to my handling editor, the incomparable Ms. Lydia Wang and also her editorial team. Her generosity and patience have been the true backbone of this book. From our first conversation to the very last line we fine-tuned, Lydia's guidance has been invaluable. She didn't just believe in this book; she believed in me. And let's be honest, that's no small feat, given that at one point, I was convinced I could barely write a coherent paragraph. Her consistent support and sharp insights turned what was once a pile of ideas and scattered thoughts into a polished, academic piece. So, Lydia, if I ever publish a sequel, I promise to keep the deadlines a little more reasonable.

Next, to my research assistants—who often had no idea they were working on this book while juggling their own studies. You know who you are, and I thank you for your patience (and for not running away screaming when I dumped piles of research via Wechat or tutorials). Yang Xiran, Sun Mingyang, Liu Chengyang, Sun Jia, Zhang Lijun, Zhang Tingya, Wang Linjue, Yin Fangqi, Wang Qian, and Cao Mengke—this book is as much almost yours as it is mine. You've poured your energy into editing, proofreading, and offering invaluable responses. I sometimes joke that I "enslave" my students, but in reality, they are the unsung associates who make me look much better than I am. The true academic rising stars—thank you for your hard work and your endless willingness to help me refine every page.

To all my students in the Global Climate Governance classes, you are my constant inspiration. There's something deeply gratifying about watching students grow from the curious minds they were when they first walked into my classroom to the climate-conscious advocates they become by the end of the course. I often learn as much from them as they do from me—if not more. You remind me why I do this work and why climate governance is, indeed, a cause worth fighting for.

A special thanks to my dear colleagues, both at home and abroad. To Prof. Dr. Yan Li of Ritsumeikan Asia Pacific University in Japan and Prof. Peter Heck of the Institute for Applied Material Flow Management (IfaS) at Trier University in Germany, Dr. LU Hongyan and Dr. YAN Jiong who have been invaluable mentors throughout these years. You have consistently challenged me to think more critically and write more profoundly. Your guidance has not only helped me grow as an academic but also as a person. To Prof. Zhao Ruixi of Hebei Normal University, Prof. Wang Wenqi, Prof. Ren Mu, Dr. Cui Yue, and Prof. Chen Xi of SIPA—your intellectual generosity and thought-provoking discussions have been instrumental in shaping my ideas and approach to this work.

I would also like to acknowledge the unwavering support of my dear colleagues and brothers, Dr. Michael Knaus and Dr. Ranahansa Sasanayake, along with other colleagues from IfaS and the IMAT Yantai program. You have been my intellectual and educational companions during some of the most challenging and stimulating phases of this project. Your insights into sustainable practices and climate diplomacy have provided the global perspective that was crucial in enriching this work.

No academic endeavor would be complete without the thankless, invisible work of funding institutions. To the Global Energy & Climate Governance Teaching and Research Platform Project of Jilin University (TS2023017), Jilin Office of Philosophy and Social Science (SKZ2023085), People's Political Consultative Conference Theory Research Center project of Jilin University (NO. 2021zx03010), and the various other sponsors who supported this journey, your backing has been the backbone of this project. Without you, the research and writing would have remained an academic dream, gathering dust on my desk.

And, of course, the most profound gratitude of all goes to my family. To my wife, Liu Zhanyan, whose understanding, patience, and love have been the bedrock on which this entire project was built. To my mother, Cong Rong, and my in-laws—thank you for always being there with love and support, even if you don't quite understand the complexities of global climate governance (and who can blame you?). To my cousins Cong Jiayuan and Jiang Ziqi and extended family who often had to put up with my climate change rants at family gatherings—your patience and willingness to engage with my passion mean more than words can express.

Lastly, to my precious daughter, Zou Yun-an. This book is dedicated to you. My work may be about climate change, but you are my daily reminder of what's at stake. Every chapter I wrote, every page I revised, was done with the hope that your future will be better, brighter, and more sustainable because of the work we do today. I hope one day you will read this book and understand why I poured my heart and soul into it. You are the reason this journey, with all its ups and downs, has been worth it.

To everyone who has helped me along the way, whether you were an intellectual collaborator, a moral supporter, or simply someone who lent an ear when I needed to vent, I thank you from the bottom of my heart. This book is as much yours as it is mine. Let's keep working together to make the world a better place—not just for us, but for generations yet to come.

Competing Interests The author declares the following as potential competing interests: This book project was funded by the Global Energy & Climate Governance Teaching and Research Platform Project of Jilin University (TS2023017), Jilin Office of Philosophy and Social Science (SKZ2023085), People's Political Consultative Conference Theory Research Center project of Jilin University (NO. 2021zx03010).

Prologue: The Unfolding Conundrum—Three Decades of Global Climate Governance

"To see what is in front of one's nose needs a constant struggle."

— George Orwell

The Unheeded Alarm: A Planet on the Brink

The story of global climate governance over the past three decades is one of a dramatic race between escalating planetary crises and a persistently laggard political response. It is a narrative punctuated by moments of collective hope and profound disillusionment, of scientific clarity mired in political obfuscation. The journey from the landmark Earth Summit in Rio de Janeiro in 1992 to the tense deliberations in Glasgow in 2021 reveals a fundamental, and perhaps deepening, schism: the chasm between the unambiguous warnings of climate science and the fractured, often self-interested, machinery of international politics.

The scientific foundation, as laid out in abundant literature, is stark and unequivocal. Since the industrial revolution, human activities—primarily the burning of fossil fuels—have pumped greenhouse gases into the atmosphere at an unprecedented rate. The data speaks a sobering language: the decade of 2011–2020 was the warmest on record, with the global average temperature already 1.1°C above pre-industrial levels (1850–1900) [1]. This warming is not an abstract statistic; it manifests in the relentless retreat of glaciers worldwide, a trend meticulously documented since the eighteenth century, and in the increasing frequency and ferocity of extreme weather events—droughts, wildfires, floods, and super-storms [2]. By 2024, the symbolic and critical 1.5°C warming threshold, enshrined in the Paris Agreement as the safer limit, had been breached, serving as a grim testament to the inadequacy of the global response [3]. This scientific reality, however, has collided with a political landscape defined by competing sovereignties, divergent economic interests, and a deep-seated struggle over fairness and historical responsibility. The core challenge of climate change is its quintessential global commons problem: the atmosphere belongs to no one and everyone, yet the costs of action are national and immediate, while the benefits are global and diffuse. This inherent tension has shaped every negotiation, every agreement, and every failure over the last thirty years.

The Rio Dawn: Forging a Framework Amidst Division

The 1992 United Nations Conference on Environment and Development in Rio de Janeiro was a watershed moment. Against a backdrop of post-Cold War geopolitical shifts and growing environmental consciousness, the international community came together to birth the United Nations Framework Convention on Climate Change (UNFCCC). Its ultimate objective was ambitious: "to achieve... stabilization of greenhouse gas concentrations in the atmosphere at a level that would prevent dangerous anthropogenic interference with the climate system" [4]. The genius, and the fundamental weakness, of the UNFCCC lay in its foundational principles. It established the bedrock concept of "Common but Differentiated Responsibilities and Respective Capabilities" (CBDR-RC). This principle formally acknowledged the historical debt of developed nations, whose centuries of industrial activity had primed the atmospheric pump, and placed the onus of leadership on them. It carved the world into Annex I (developed) and non-Annex I (developing) countries, creating a binary that would define climate politics for decades [5]. While a necessary political compromise to secure universal participation, this division sowed the seeds for future conflict, as the economic rise of nations like China and India would later strain its simple dichotomy.

The Convention also established the Conference of the Parties (COP) as its supreme governing body, launching an annual ritual of international diplomacy that continues to this day. It was a moment of optimism, a belief that a framework for cooperation could tame a global problem. Yet, the UNFCCC was just that—a framework. It set no binding targets, offering only a vessel into which concrete commitments would hopefully be poured.

The Kyoto Protocol: The "Top-Down" Experiment and Its Discontents

The first major attempt to fill this vessel was the 1997 Kyoto Protocol. It represented the high-water mark of the "top-down" approach to climate governance. In a significant leap, it assigned legally binding emission-reduction targets exclusively to Annex I countries, requiring them to cut their collective emissions by an average of 5.2% below 1990 levels by 2012 [6].

Kyoto's innovation was its introduction of market-based "flexibility mechanisms," most notably the "Clean Development Mechanism (CDM)." The CDM allowed developed countries to invest in emission-reduction projects in developing nations and receive credit for the cuts. In theory, it was a win-win: cost-effective mitigation for the North and sustainable development finance for the South [7]. However, in practice, its benefits were unevenly distributed. Over 80% of CDM projects were concentrated in China and India, while the poorest nations, particularly in Africa, were largely bypassed [8]. Furthermore, technology transfer was often superficial, involving equipment rather than core knowledge, limiting its long-term developmental impact [9]. The Kyoto model, however, was structurally fragile. Its fatal flaw was exposed when the United States, then the world's largest emitter, signed but never ratified the protocol, citing economic concerns and the lack of commitments from major developing economies. This move crippled the protocol's global effectiveness and signaled the profound difficulty of reconciling national economic competitiveness with global ecological imperatives. The "top-down" model, reliant on legally mandated targets for a subset of countries, proved politically unsustainable in a world of shifting economic power and entrenched "free-rider" incentives.

The Paris Agreement: A Paradigm Shift to "Hybrid" Governance

The collapse of the "top-down" model culminated in the acrimonious failure of the 2009 Copenhagen Summit (COP15). The ensuing search for a new approach led to the landmark "Paris Agreement" in 2015. Paris represented a fundamental paradigm shift, moving from a "Kyoto-style" prescribed regime to a "hybrid" model of governance. At its heart lies the mechanism of "Nationally Determined Contributions (NDCs)." This "bottom-up" approach allows each country to determine its own climate actions based on its national circumstances, capabilities, and assessments. This ensured near-universal participation, as it respected national sovereignty and political feasibility. To counter the inherent risk of low ambition, the Agreement established a "ratchet mechanism": a five-yearly "Global Stocktake" to assess collective progress and compel parties to progressively enhance their NDCs [10]. The Paris Agreement also skillfully, and contentiously, reframed the CBDR-RC principle. It moved away from the rigid Annex I/non-Annex I binary, instead calling for action "in light of different national circumstances." This semantic shift was crucial for bringing the United States and emerging economies on board, but it was perceived by many developing countries as a dilution of historical responsibility and a concession to developed nations [11].

Despite its architectural elegance, the Paris model has a critical weakness: the voluntary nature of the NDCs. There is no enforcement mechanism to ensure countries meet their pledges or increase their ambition sufficiently. The result, as the 2023 Global Stocktake confirmed, is a yawning "ambition gap"—even if all current NDCs are fully implemented, they would lead to a devastating 2.5°C of warming, far exceeding the Agreement's goals [12]. The "hybrid" model successfully brokered universal participation, but at the cost of collective adequacy.

The Fractured Stage: Key Actors in the Climate Arena

The evolution of climate governance cannot be understood without examining the key actors whose interests and ideologies have shaped its trajectory. The negotiation blocs that emerged from the UNFCCC process are microcosms of the global political economy of climate change.

- The European Union (EU): Often cast as the "climate leader," the EU has consistently pushed for ambitious targets and developed the world's first major carbon market, the EU Emissions Trading System (EU ETS). Its leadership stems from a combination of public pressure, strategic interest in clean technology, and a political identity built on multilateralism [13]. However, its influence has been challenged by internal divisions, particularly between Western European front-runners and Eastern European members more reliant on fossil fuels.
- The Umbrella Group: This loose coalition of non-EU developed countries, including the United States, Japan, Canada, and Australia, has traditionally emphasized economic costs and competitiveness. Their stance has often been one of caution, resisting stringent, legally binding cuts and insisting on meaningful participation from major emerging economies [14]. The group's coherence has been repeatedly tested by the volatile climate policies of its most powerful member, the United States, which has oscillated between engagement and withdrawal under different administrations.
- The Group of 77 and China (G77+China): This large and diverse coalition has been the primary guardian of the CBDR-RC principle. Its members argue that developed nations must bear the brunt of mitigation and provide the necessary finance and technology to the developing world. However, the bloc is far from monolithic. It contains deep internal fissures between:
- Rapidly Emerging Economies (e.g., China, India, Brazil, South Africa): These nations fiercely defend their "right to develop" and resist any dilution of differentiation that would impose a premature mitigation burden on their economies.
- The Alliance of Small Island States (AOSIS): For these countries, climate change is an existential threat. They have been the most vocal advocates for the 1.5°C target and the most critical of large emitters, both developed and developing. Their influence derives not from economic power, but from the potent moral authority of their plight [15].
- The Least Developed Countries (LDCs): Focused on adaptation and financial support, they often find their urgent needs overshadowed by mitigation debates among larger powers.
- Oil-Producing Nations (e.g., OPEC members): These states have frequently acted as a "veto coalition," resisting language on fossil fuel phase-outs and emphasizing

the economic disruption stringent climate policies would cause to their resource-dependent economies [16].

This complex tapestry of interests ensures that every climate negotiation is a multi-level game of chess, where North-South divides are constantly cross-cut by sub-group alliances and competing vulnerabilities.

Beyond the State: The Rise of Non-State Actors and Systemic Challenges

The past three decades has also witnessed the gradual erosion of the state-centric model of climate governance. A "polycentric" or "complex governance" system has emerged, involving cities, regions, corporations, financial institutions, and civil society groups [17]. The "Fridays for Future" movement, ignited by Greta Thunberg, exemplifies how non-state actors can shape public discourse and exert moral pressure on governments [18]. Similarly, corporate net-zero pledges and city-level climate action initiatives have proliferated, creating pockets of progress even in the face of national-level inertia. Yet, as the attached materials critically note, these actors often remain on the margins of formal decision-making processes within the UNFCCC. Their potential for resource mobilization, innovation, and filling governance gaps is still underestimated and under-institutionalized by the state-dominated regime [19].

Furthermore, the governance challenge is compounded by deeper systemic lock-ins. The global economy remains structurally dependent on fossil fuels. A complex interplay of political, economic, and technological forces across interconnected subsystems—national economies, global markets, urban systems, and daily life—reinforces a high-carbon development pathway. Breaking this "carbon lock-in" requires a fundamental transformation that current governance arrangements are not yet equipped to deliver.

At the Crossroads of Civilization

Thirty years after Rio, global climate governance stands at a precipice. The initial framework-building phase is long over. The experimental phase of "top-down" targets and "bottom-up" pledges has revealed both the limits of coercion and the perils of voluntarism. The science has progressed from cautious prediction to desperate warning. The political debates have evolved, but the core dilemmas remain: equity versus effectiveness, sovereignty versus solidarity, and short-term cost versus long-term survival. The "carbon power" dynamics are shifting, with new battlegrounds emerging over green technology dominance, carbon border taxes, and the terms of climate finance. In this new arena, claims for "climate justice" are often subsumed by geopolitical calculation, and the voices of the most vulnerable are systematically marginalized.

The path forward is fraught, but not invisible. It requires a cognitive reframing of the problem itself. Climate change is not merely an environmental issue to be "solved" by technical fixes within the existing economic paradigm. It is a symptom of a deeper civilizational conflict between an infinite growth model and a finite planetary system. Future governance must therefore move beyond tinkering with emission allocations and embrace a more transformative agenda. This includes:

- Reinforcing the Hybrid Model: Strengthening the Paris Agreement's transparency framework and ratchet mechanism to create a more potent cycle of accountability and ambition.
- Mainstreaming Justice: Operationalizing climate justice through robust and accessible finance for "loss and damage," equitable technology transfer, and a just transition that leaves no one behind.
- Embracing Polycentricity: Formally integrating the agency of cities, businesses, and investors into the heart of the implementation architecture, turning their dispersed actions into a coordinated global force.
- Confronting Carbon Lock-in: Designing policies that actively dismantle fossil fuel subsidies, redirect financial flows toward green infrastructure, and foster the socio-technical innovations needed for a post-carbon world.

As one poignant metaphor in the materials suggests, humanity is on a supertanker heading for reefs, while the crew debates below deck [20]. The voyage of global climate governance over the past thirty years has been this debate—necessary, contentious, and agonizingly slow. The chapters that follow will delve into the specific roles and records of the key members of this crew—the major powers, the negotiating blocs, and the international institutions. But as this prologue has argued, the fundamental lesson of the past three decades is that while steering the ship is difficult and the precise course uncertain, the imperative to collectively seize the helm is no longer a matter of choice, but of survival. The journey from Rio to Glasgow has brought us to the heart of the storm; the journey ahead will determine whether our civilizational vessel can be steered toward safer waters.

References

1. Met Office Hadley Centre. Global mean temperature difference from 1850–1900. Climate Dashboard. (Data referenced in original Chinese source material)
2. Dessler A, Parson EA (2012) The science and politics of global climate change (trans: from Chinese). China Environmental Science Press
3. World Meteorological Organization (WMO) (2024) 2023 officially the hottest year on record. [Press release summary from original material]
4. United Nations (1992) United Nations framework convention on climate change. Article 2
5. Rajamani L (2016) The 2015 Paris agreement: interplay between hard, soft and non-obligations 28(2)

6. United Nations (1998) Kyoto protocol to the United Nations framework convention on climate change
7. United Nations (1998) Kyoto protocol. Article 12
8. Carbon Markets Almanac (2015) Project distribution of the clean development mechanism. (Data referenced in original English source material)
9. Center for Climate and Energy Solutions (C2ES) (2011) The clean development mechanism: a review of the first international offset program
10. United Nations (2015) Paris agreement. Articles 4 & 14
11. Brunnee J, Streck C (2013) The UNFCCC as a negotiation forum: towards common but more differentiated responsibilities. Clim Policy 13(5)
12. UNFCCC (2023) Technical dialogue of the first global stocktake. Synthesis Report
13. Szulecki K et al. (2016) Shaping the 'Energy Union': between national positions and governance innovation in EU energy and climate policy. Clim Policy 16(5)
14. Morgan G (2001) The collapse of the kyoto protocol and the struggle to slow global warming. Princeton University Press
15. Betzold C (2010) 'Borrowing' power to influence international negotiations: AOSIS in the climate change cegime, 1990–1997. Polit 30(3)
16. OPEC Bulletin (2007) OPEC's Perspective on climate change. (Referenced in original English source material)
17. Kuyper J, Schroeder H, Linnér B-O (2018) The evolution of the UNFCCC. Annu Rev Environ Resour 43
18. Thunberg G (2019) No one Is too small to make a difference. Penguin Books
19. Andonova LB, Betsill MM, Bulkeley H (2009) Transnational climate governance. Glob Environ Polit 9(2)
20. Dessler A, Parson EA (2012) The science and politics of global climate change. China Environmental Science Press, p 68. (Metaphor of the supertanker, paraphrased from the original Chinese text)

Contents

About the Author

Dr. Xiaolong Zou is an Associate Professor and Ph.D. Supervisor at the School of International and Public Affairs (SIPA) at Jilin University, China. He is also the Head of the Department of Global Governance & International Organization at SIPA. With over a decade of teaching and research experience, Dr. Zou has become an active voice in the field of global climate governance and international environmental policy.

Dr. Zou holds a Ph.D. in Asia Pacific Studies from Ritsumeikan Asia Pacific University in Japan, and has worked extensively on global energy and climate governance, focusing on climate diplomacy, sustainable development, and the political dynamics of the Global South. He has published widely in prestigious academic journals indexed by SSCI, A&HCI, and SCI, including Transport Policy, Journal of Contemporary Asia, South East Asia Research, and Energy. His research has secured funding for over ten provincial and ministerial-level projects, and he has contributed to shaping climate policy through his work as an expert reviewer for the China Scholarship Council (CSC) and as a Board Member of the China United Nations Association.

In addition to his academic achievements, Dr. Zou has served as a Visiting Professor at Ritsumeikan Asia Pacific University in Japan and a Visiting Research Fellow at the Institute for Applied Material Flow Management (IfaS) at Trier University of Applied Sciences in Germany. These international experiences have given him a unique, global perspective on the challenges and opportunities in international climate diplomacy.

Beyond his academic work, Dr. Zou is deeply committed to educating the next generation of scholars and policymakers. His courses on global climate governance and environmental politics have inspired countless students to engage critically with the pressing issues of our time. Through his work, both in the classroom and in his publications, Dr. Zou continues to push the boundaries of scholarship and policy, aiming to influence both global climate governance and the real-world actions needed to address the climate crisis.

When he is not teaching or researching, Dr. Zou enjoys spending time with his family, hiking, and reflecting on the intersection of nature, politics, and society. He is driven by the belief that the world can change if we all take responsibility, and that sustainable futures are built through education, collaboration, and innovative thinking.

Chapter 1
The Role of International Organizations in Shaping Global Climate Governance

The architecture of global climate governance, constructed over the three decades from the Rio Earth Summit to the Glasgow Climate Pact, is fundamentally a story of international organizations. These entities—each with distinct mandates, powers, and constituencies—have been the primary architects, engineers, and custodians of the global response to the climate crisis. They have transformed a complex scientific phenomenon into a sustained subject of high politics, forged fragile agreements among sovereign states, and generated the knowledge necessary for action. This chapter examines the pivotal roles of four such organizations: the United Nations (UN), the Intergovernmental Panel on Climate Change (IPCC), the International Energy Agency (IEA), and the United Nations Environment Programme (UNEP). It argues that while their individual contributions are indispensable—from the UN's unparalleled convening power and the IPCC's scientific authority to the IEA's technical expertise and UNEP's targeted assessments—their collective efficacy is persistently challenged by structural fragmentation, political constraints, and a widening gap between global ambition and implementation. By analyzing their evolution, strengths, and vulnerabilities, this chapter illuminates not only the central pillars of the existing governance regime but also the necessary pathways for its reinforcement in an era of escalating climate impacts and geopolitical fragmentation.

1.1 The United Nations: Its Role, Challenges, and Pathways in Global Climate Governance

The United Nations stands as the most authoritative and influential institution in global climate governance. A clear understanding of its role, the challenges it confronts, and potential pathways for enhancement is crucial for maximizing its efficacy. This section begins by outlining the organizational foundation of the UN's participation in climate governance. It then employs Robert Keohane's 3Cs analytical

X. Zou, *Thirty Years of Global Climate Governance*, Energy, Climate Change & Sustainability in the Global South, https://doi.org/10.1007/978-981-95-8457-4_1

framework for effective global environmental governance to dissect the UN's contributions: elevating governmental concern, facilitating contractarian environments, and enhancing implementation capacity. Finally, it identifies the challenges the UN faces—including a fragmented governance structure, insufficient capacity, and the risk of marginalization—and proposes responses centered on institutional reform, capacity building, and strategic repositioning for more effective climate action.

1.1.1 The Organizational Foundation of UN Climate Governance

The UN's involvement in climate governance is facilitated through a tripartite institutional framework. The first category comprises dedicated agencies established specifically to address climate change, including the Intergovernmental Panel on Climate Change (IPCC), the Intergovernmental Negotiating Committee (INC), and the Secretariat of the United Nations Framework Convention on Climate Change (UNFCCC). The second category consists of permanent agencies with broader, comprehensive mandates within the UN system, such as the Economic and Social Council (ECOSOC), the United Nations Environment Programme (UNEP), the United Nations Development Programme (UNDP), the World Meteorological Organization (WMO), and specialized bodies like the International Maritime Organization (IMO). The third category encompasses auxiliary and specialized agencies, including the Global Environment Facility (GEF) and the World Climate Programme Coordination Committee (CCWCP). This extensive, albeit sometimes overlapping, network forms the institutional backbone of the UN's climate efforts [1].

1.1.2 The Role of the UN in Climate Governance: A 3Cs Framework Analysis

To analyze the role of the UN in climate governance, this section adapts the 3Cs framework proposed by Robert Keohane and others, which posits that effective global environmental governance requires [2]:

- **Concern**: A high level of governmental attention to the environmental issue.
- **Contractual Environment**: A negotiating environment that mitigates fears of free-riding and fraud, enabling reliable commitments, adherence to common rules, and low-cost monitoring for reciprocal cooperation.
- **Capacity**: The political and administrative capability of states to make domestic adjustments to comply with international norms and principles.

Climate change, while distinct in scale and political sensitivity, shares fundamental characteristics with traditional environmental problems—such as their public goods nature, potential irreversibility, cognitive lags, and high mitigation costs [3]. These commonalities allow for the application of the 3Cs framework. However, climate change's heightened political stakes, higher scientific complexity, and the unique dimension of 'adaptation' necessitate specific UN roles: integrating divergent national interests, enhancing scientific literacy among governments and the public, and promoting adaptation policies alongside mitigation [4].

Based on this, the UN's role in climate governance manifests in three key areas:

a. **Elevating Concern and Setting the Global Agenda**

The UN has systematically elevated climate change from a scientific concern to a cornerstone of global politics. A series of pivotal actions—from the initial climate risk warnings in the 1972 UN Human Environment Conference's Action Plan and the 1979 First World Climate Conference, to the politically significant 1985 Villach Conference and the 1987 Brundtland Report—culminated in the establishment of the IPCC in 1988. The IPCC's successive assessment reports have been instrumental in building a robust scientific consensus on human-induced warming. This trajectory led the UN General Assembly to initiate negotiations, directly resulting in the adoption of the UNFCCC in 1992, which formally inaugurated the era of collaborative global climate governance.

b. **Fostering a Conducive Contractual Environment**

The UN creates a framework for negotiation and cooperation essential for overcoming collective action problems.

- **Providing a Forum for Negotiation**: The primary venue is the Conference of the Parties (COP) under the UNFCCC, the convention's supreme decision-making body. Supported by technical and logistical assistance from bodies like the UNFCCC Secretariat, WMO, and UNEP, COPs facilitate multilateral negotiations on mitigation, adaptation, finance, and technology transfer.
- **Institutionalizing Long-term Governance Mechanisms**: The UN has championed and solidified the "Convention-Protocol" model of negotiations, persistently advancing emission reduction targets and timelines. It has upheld the critical principle of "common but differentiated responsibilities," navigating complex geopolitical landscapes. Despite setbacks, such as the US rejection of the Kyoto Protocol, the UN provided a platform for salvaging agreements like the Bonn Political Agreement and the Marrakech Accords, ensuring the Protocol's eventual entry into force [5].
- **Addressing the 'Collective Action Dilemma':** The UN has facilitated voluntary cooperation mechanisms to lower compliance costs and foster sustainable development. The Kyoto Protocol established a 1.0 version of such mechanisms, including International Emissions Trading (ET), Joint Implementation (JI), and the Clean Development Mechanism (CDM). The Paris Agreement ushered in a 2.0 version, based on a "bottom-up" model where parties submit Nationally Determined Contributions (NDCs), which are then subject to international assessment and updating.

- **Enhancing National Capacity for Implementation**

UN agencies play a vital role in building the capacity of states to fulfill their climate commitments. UNEP serves as a prime example. It implements environmental projects worldwide; to date, UNEP has overseen 788 projects across 149 countries, with a total value of $269 million [6]. Furthermore, according to its 2022 Programme Performance Report, 76% of its indicators fully met their targets, with all climate actions fully achieving their objectives [7]. UNEP also actively supports countries in developing and updating national adaptation plans, strategies, and policies. As of August 2023, five-sixths of UNFCCC parties had at least one such national instrument in place [8].

1.1.3 Challenges Confronting the UN in Climate Governance

a. **Fragmented Governance Structure**

The UN system is often characterized by institutional bloating and overlapping mandates. In climate governance, numerous entities beyond the core of UNEP, WMO, and IPCC—such as the World Bank Group and UNDP—engage in work that is frequently duplicative. This fragmentation leads to resource waste and diminished operational efficiency.

b. **Insufficient Governance Capacity**

The UN's practical effectiveness is hampered by three structural capacity deficits:

- **Limited Coordination and Integration**: As a multilateral platform, the UN struggles to reconcile fundamental divergences in national interests. During negotiating impasses, it often becomes an arena for great power politics rather than an active mediator, as evidenced by the Copenhagen conference where final outcomes relied on bilateral deals between major economies.
- **Weak Monitoring and Enforcement**: The lack of a robust compliance mechanism with meaningful penalties means that member state violations can only be addressed through non-binding dispute resolution procedures, severely undermining the force of international agreements.
- **Persistent Financial Gaps**: The chasm between needs and resources is staggering. UNEP's 2023 Adaptation Gap Report indicates that developing countries require $387 billion annually for adaptation, while international public finance flows stood at only $21 billion in 2021, creating a gap of $194-$366 billion [9]. Compounding this, the UN Environment Fund, a core funding source, remains under-subscribed, with only 59 of 193 member states contributing as of October 2023 [10].
- **Trend Toward Marginalization**

The UN's central role in climate governance has shown signs of erosion. Vertically, escalating conflicts between sovereign states have exposed the UN's limited ability to transcend disputes and coordinate interests effectively, gradually weakening its leadership. Horizontally, the UN faces competition from alternative governance mechanisms like the Asia–Pacific Partnership on Clean Development and Climate, as well as climate initiatives within the G7, G20, and APEC. These smaller, often like-minded forums can sometimes achieve policy consensus more readily, challenging the UN's primacy [11].

1.2 The Intergovernmental Panel on Climate Change (IPCC): Scientific Backbone of Global Climate Governance

Amidst intensifying global warming and accelerating glacial melt, the Intergovernmental Panel on Climate Change (IPCC) stands as a pivotal scientific body in the climate response. This section systematically examines the IPCC's functions within global climate governance, its inherent vulnerabilities, and potential pathways for improvement.

1.2.1 An Overview of the IPCC

Established in 1988 by the World Meteorological Organization (WMO) [12] and the United Nations Environment Programme (UNEP) [13], the IPCC was created to address the threats posed by global climate change to human existence and development. Its mandate is to assess the world's best available scientific, technical, and socio-economic information on climate change on a comprehensive, objective, open, and transparent basis. It is open to all member states of the UN and WMO, providing assessments and advice to inform national climate policies.

The IPCC conducts its assessments through three Working Groups and a Task Force: Working Group I focuses on the physical science basis of climate change; Working Group II assesses impacts, adaptation, and vulnerability; Working Group III evaluates climate change mitigation; and the Task Force is responsible for the National Greenhouse Gas Inventories Programme. As the paramount scientific authority, the IPCC has published six comprehensive Assessment Reports, which have become the cornerstone for global climate policymaking. Its work has not only solidified the scientific foundation for action but has also consistently underscored the profound urgency of the climate crisis—highlighting that delayed action poses an existential threat to humanity and ecological stability.

1.2.2 The Role of the IPCC in Global Climate Governance

a. **Providing Scientific Assessment and Early Warning**

The IPCC, adhering to principles of scientific rigor, objectivity, and transparency, has systematically synthesized global climate research. Its six Assessment Reports have provided the critical scientific impetus for international climate governance:

- **The First Assessment Report** (1990) confirmed the reality of global warming, directly catalyzing the negotiation of the UNFCCC.
- **The Second Assessment Report** (1995) strengthened the evidence for a "discernible human influence" on the global climate, paving the way for the Kyoto Protocol.
- **The Third Assessment Report** (2001) concluded that human influence was likely (66% probability) responsible for observed warming.
- **The Fourth Assessment Report** (2007) declared the warming of the climate system "unequivocal," with a very high confidence (90% probability) that it was predominantly anthropogenic.
- **The Fifth Assessment Report** (2014) increased the certainty of human causation to over 95%, forming the scientific bedrock for the Paris Agreement.
- **The Sixth Assessment Report** (2021–2023) revealed that global surface temperature has already risen by 1.1 °C and warned that, under current emissions pathways, the 1.5 °C threshold could be breached by the early 2030s, issuing an urgent call for intensified mitigation.

This progression of reports has progressively fortified the scientific consensus on human-driven climate change, providing a continuously updated evidence base for global negotiations and agreement formation [14]. The evolution of these reports demonstrates the IPCC's significant contribution to shaping national climate policies and the broader governance architecture [15].

b. **Influencing National Climate Policy Formulation**

China's climate policy trajectory serves as a compelling case study. As the world's largest emitter, whose share of global emissions has risen from 15% to nearly one-third over two decades, China has been responsive to the urgency highlighted in IPCC reports. This is evidenced by its commitment to achieving carbon neutrality before 2060, a pledge directly informed by the IPCC's scientific assessments [16]. This commitment, from the world's largest investor in renewable energy, is projected to lower global warming by 0.2–0.3 °C and accelerate the global carbon neutrality timeline by 5–10 years, demonstrating the tangible policy impact of the IPCC's work [17].

c. **Promoting International Cooperation and Institutional Refinement**

IPCC assessment reports provide direct scientific guidance for international policy. The Fifth Assessment Report (AR5) was particularly instrumental in shaping the Paris Agreement. It provided a comprehensive evaluation of different greenhouse gas concentration pathways and their implications for global warming, which informed the Agreement's long-term temperature goal. It was central to the scientific discourse that led to the enshrinement of the 1.5 °C target [18]. This threshold is critically important as it represents a key defense line against severe and often irreversible climate impacts that would become far more pronounced at 2 °C of warming, thereby significantly reducing long-term risks to societies and ecosystems.

d. **Vulnerabilities and Shortcomings in IPCC Operations**

As a hybrid scientific and intergovernmental body, the IPCC inevitably exhibits certain vulnerabilities in its internal cohesion, external adaptability, and research depth. Identifying these issues is a prerequisite for enhancing its efficacy [19].

Insufficient Internal Cohesion and Standardization

The IPCC's structure presents inherent challenges. Its members, drawn from diverse national and disciplinary backgrounds, lack a shared cultural foundation and sustained communication channels. When faced with ambiguous scientific evidence, this can lead to diverging viewpoints that hamper assessment efficiency. The protracted assessment cycles, coupled with limited collaborative time for authors, can constrain deep deliberation and consensus-building. Such mechanistic flaws have occasionally resulted in errors, most notably the 2007 "Glaciergate" incident, where an unsubstantiated claim about the imminent disappearance of Himalayan glaciers damaged the Panel's scientific credibility and revealed lapses in review procedures.

The Rise of Anti-Globalization and International NGOs

The IPCC faces a dual challenge from resurgent anti-globalization sentiments and the growing influence of non-governmental organizations. As a scientific advisory body, the IPCC lacks effective mechanisms to mediate fundamental conflicts of national interest, limiting its ability to foster deep international cooperation. Its policy influence has also proven vulnerable to abrupt political shifts, such as the UK's Brexit referendum and the U.S. withdrawal from the Paris Agreement under the Trump administration, exposing its fragility in the face of political volatility. Concurrently, the rise of influential NGOs like the World Wide Fund for Nature (WWF) [20], while contributing positively to climate action, also creates a more crowded and competitive landscape, potentially diluting the IPCC's unique voice and support.

Lack of Deep Exploration into Socio-Political Dimensions

The IPCC's policy orientation tends to prioritize consensus and avoid conflict. Consequently, its assessments predominantly focus on climate science and technical mitigation pathways, while giving insufficient systematic attention to the deeper socio-political equity issues exacerbated by climate change, such as global wealth disparity and the unfair distribution of responsibilities and resources between developed and developing countries. Furthermore, the highly technical and cautious language of its reports presents a high barrier to public comprehension, limiting their broader societal reach and impeding their potential to mobilize wider public engagement in climate action.

e. **Reform Measures**

To address these operational vulnerabilities and enhance the IPCC's effectiveness, the following reforms are proposed:

Enhancing Internal Cohesion and Standardization

Strengthening communication and collaboration among authors is key. The IPCC could further diversify its author base by incorporating insights from a wider range of disciplines, including the social sciences and humanities, to enrich the breadth and depth of its reports. Efforts should be made to safeguard the scientific process from the veto power of self-interested government reviewers during the approval of Summaries for Policymakers, thereby preserving the reports' scientific integrity and accuracy from overt political bargaining [21]. Implementing more robust internal and external review and evaluation mechanisms could also help ensure the rigor and credibility of its work.

Strengthening International Cooperation and Coordinating Conflicts

Confronting anti-globalization trends and a diversified NGO landscape, the IPCC must work to consolidate a global cognitive consensus on climate change, making "jointly solving the global climate crisis" a universal conviction among nations and their citizens [22]. To this end, the IPCC should not only serve as an authoritative platform for disseminating climate science but also proactively assume a role in mediating disputes between nations. It must deepen international cooperation, guiding governments to jointly deliberate on responses and share governance responsibilities, thereby reinforcing the development of multilateralism through collective climate action.

Focusing Climate Policy on Deeper Issues

The IPCC should broaden its perspective beyond pure climate science. It must integrate issues of social equity, fair responsibility-sharing, and just transition into its assessment framework, ensuring climate policies are not only scientifically sound but also socially responsive and equitable. Secondly, it should optimize its communication strategy by adopting more accessible language and illustrative case studies to enhance the readability of its reports for non-specialist audiences. By improving its policy relevance and societal resonance, the IPCC can more effectively serve the global climate governance process.

1.3 The International Energy Agency (IEA): Evolution and Expanding Mandate in Climate Governance

Established in 1974 by the Organisation for Economic Co-operation and Development (OECD) in response to the oil crisis, the International Energy Agency (IEA) initially focused on coordinating measures to prevent severe oil supply disruptions and providing authoritative statistics on the international oil market. Over time, its mandate has significantly broadened. The organization's mission is now concentrated on promoting "3E" energy policies: Energy security, Economic development, and Environmental protection [23]. This evolution reflects a strategic pivot from a narrow focus on oil security to a comprehensive approach encompassing global energy transitions and climate mitigation.

1.3.1 Membership, Structure, and Governance

The IEA currently comprises 31 full member countries and has engaged 13 associate members. Accession to full membership is a rigorous process, requiring an applicant country to first be an OECD member. Furthermore, it must meet stringent conditions, including holding oil stocks equivalent to 90 days of the previous year's net imports, having a national demand restraint program to reduce oil consumption by up to 10% in an emergency, and establishing a legal and organizational framework for coordinated crisis response [24]. This high bar for membership is designed to ensure the coherence and effectiveness of the IEA's core collective action mechanism: the coordinated release of strategic petroleum reserves during major global supply disruptions, a system that has been activated during events such as the 1991 Gulf War and the 2022 Russia-Ukraine conflict [25].

The IEA's structure is built around four principal organs: the Governing Board, Standing Groups, Committees, and the Secretariat [26]. The Governing Board, which meets at both ministerial and official levels, serves as the apex decision-making body. It operates largely through consensus, with many of its proceedings being confidential. A notable feature of its 1990s governance was a pattern of "institutional adjustment with limited delegation of authority," where core decision-making power remained centralized while auxiliary tasks were selectively assigned to subordinate bodies [27]. The Secretariat, led by the Executive Director, undertakes the day-to-day operations. The Executive Director wields significant managerial and representational authority, including negotiating policy agreements with non-member countries and managing the agency's budget with considerable flexibility [28].

1.3.2 The IEA's Role in Climate and Energy Governance

The IEA's mission—"to shape a secure and sustainable energy future for all" [29]—is operationalized through its dual role as a key player in climate governance and energy security, primarily by leading international cooperation and providing authoritative data, analysis, and policy solutions.

Leading a New Era of International Energy Cooperation

The IEA fosters collaboration through several key channels:

- Technology Collaboration: The IEA's Technology Collaboration Programme (TCP) provides a foundational framework for international public and private research partnerships. It supports networks of experts dedicated to advancing research, development, and the commercialization of energy technologies across sectors like buildings, electricity, transport, and renewables, directly supporting the goals of energy security and environmental protection [30].
- Multilateral Engagement: The IEA serves as a central platform for global energy dialogue, working closely with international fora such as the G20, APEC, and UNFCCC. It hosts major events like the Clean Energy Ministerial and engages in joint studies and policy coordination with organizations like OPEC to advance global energy security and the low-carbon transition [31].
- Global Capacity Building: The agency has established a comprehensive global training network covering energy statistics, modeling, and energy efficiency. This system, which has expanded from its initial focus on oil emergencies to include renewables and other emerging areas, is crucial for building institutional capacity worldwide [31].

Providing Authoritative Analysis and Policy Solutions

The IEA has built an unparalleled reputation as a source of objective energy information, which is critical for informed policymaking in climate and energy.

- Understanding the Global Energy System: The agency regularly publishes flagship reports such as the World Energy Outlook and the World Energy Investment Report. These publications, prepared by in-house specialists and energy modelers, provide critical insights into current energy trends and future scenarios, guiding both international policy and the IEA's own strategic direction.
- Developing Authoritative Data and Visualizations: A core function of the IEA is the systematic collection, analysis, and dissemination of energy data. Its comprehensive datasets and interactive visualizations are indispensable tools for tracking global energy consumption, forecasting future trends, and assessing progress towards climate and energy goals.
- Engaging in Global Dialogue: The IEA actively participates in the global discourse on energy and climate governance. Through published commentary, articles, and

direct engagement in policy debates, it voices its expert opinions and recommendations, ensuring its analysis directly influences national and international decision-making processes.

1.3.3 Conclusion on the IEA

Since its inception, the IEA has successfully transformed from a specialized agency focused on oil supply security into a leading authority on energy security and climate governance. This transformation has been underpinned by its development of a robust data infrastructure and its active multilateral engagement, which have collectively built its professional credibility. However, in an era marked by resurgent anti-globalization sentiments and geopolitical tensions, the influence of international organizations like the IEA faces significant headwinds. While it provides an essential platform for dialogue and a normative framework for its members, the frequent undermining of international agreements and the breaking of commitments in recent years highlight the pervasive challenges confronting multilateral institutions [32].

1.4 The United Nations Environment Programme (UNEP): Analysis and Assessment Through the Lens of Food Loss and Waste

The United Nations Environment Programme (UNEP), established in 1972 and headquartered in Nairobi, Kenya, holds a unique position as one of the few UN agencies based in a developing country. Its membership is open to all UN states, and its mission is "to provide leadership and encourage partnership in caring for the environment by inspiring, informing, and enabling nations and peoples to improve their quality of life without compromising that of future generations." Its core functions include promoting international environmental cooperation, providing policy guidance within the UN system, conducting global environmental assessments, and facilitating the exchange of environmental knowledge [33]. This section will utilize the issue of food waste as a critical entry point to examine UNEP's specific role in scientific assessment, policy coordination, and global advocacy within the context of climate governance.

The intrinsic link between food waste and climate change is unequivocal. Wasted food, when left to decompose in landfills, generates significant amounts of greenhouse gases, particularly methane, a potent short-lived climate pollutant. The scale of the problem is monumental, with approximately one-third of all food produced for human consumption lost or wasted globally. The carbon footprint of global food waste is so substantial that, if it were a country, it would rank as the third-largest emitter, following only the United States and China [34]. The climate impact of the food system is, however, not limited to the waste phase. Emissions occur throughout the entire lifecycle—from land-use change for agriculture, production, and processing to transportation and packaging. A comprehensive study published in Nature Food estimated that global food systems are responsible for about 17.3 billion metric tons of CO2 equivalent per year, with animal-based foods contributing disproportionately (57%) compared to plant-based foods (29%) [35]. This underscores that addressing food waste is not only a matter of ethical consumption but a critical climate mitigation strategy.

1.4.1 Focusing on UNEP's Functions Through the Food Waste Index Report

The 2030 Agenda for Sustainable Development, adopted in 2015, explicitly addresses this issue under Sustainable Development Goal (SDG) 12, which calls for sustainable consumption and production patterns. Specifically, Target 12.3 aims to "halve per capita global food waste at the retail and consumer levels and reduce food losses along production and supply chains, including post-harvest losses" by 2030 [36]. To support the achievement of this target, UNEP published the Food Waste Index Report in 2021. This report represents a cornerstone of UNEP's capacity-building and scientific assessment functions, aiming to provide a standardized measurement methodology for countries to track their progress.

A pivotal finding of the report challenges a common assumption: that food waste is predominantly a problem of affluent nations. The data reveals that household food waste levels in several lower-middle-income countries are comparable to, and in some cases exceed, those in high-income countries [37–39]. This counterintuitive phenomenon can be attributed to structural and behavioral factors. In many developing countries, consumers may more frequently purchase lower-cost, lower-quality food items that contain a higher proportion of inedible parts or spoil more quickly. Furthermore, a greater reliance on household cooking (as opposed to dining out) increases the opportunities for waste to occur at the consumption stage. Crucially, the data shows no strong correlation between national GDP and per capita household food waste, indicating that action is equally critical across all income levels [40, 41].

Beyond providing a global snapshot, the report's primary function is to establish a robust methodological framework. UNEP outlines suitable measurement methods for different sectors of the food system—such as direct measurement, waste composition analysis, and mass balance—and provides a clear reporting pathway through the UN Statistical Division/UNEP Environment Statistics Questionnaire [42, 43]. This systematization enables countries to generate robust, comparable data, which is the foundation for effective national strategies. By coordinating this data collection and publishing periodic global reports, UNEP plays an indispensable role in holding the world accountable to SDG 12.3, fostering meaningful international comparison, and guiding targeted action [44].

1.5 Conclusion

Tackling food loss and waste offers multiple benefits, a truth that is especially salient in a world grappling with persistent hunger and rising food prices. Through scientific assessment (the Food Waste Index Report), policy coordination (establishing SDG 12.3 reporting mechanisms), and global advocacy, UNEP has demonstrated a critical and effective role in this specific domain of climate governance. However, a broader view reveals a sobering reality: despite these targeted efforts, the global environmental crisis continues to deepen. This should not be simplistically attributed to a failure of UNEP. The core challenge lies in the insufficient implementation of its scientifically-grounded recommendations by member states. As Andrew Steer aptly noted, assessing UNEP's success yields a complex answer: the organization has, within its constrained mandate, often done all that could be expected of it [45]. This judgment acknowledges UNEP's vital contributions while hinting at the profound structural limitations of global environmental governance, where the translation of knowledge into collective, transformative action remains the greatest hurdle.

References

1. Wu Z, Xu F (2016) The United Nations and global climate governance. Soc Sci Dig (01):22–24
2. Zhang H (2007) The United Nations and international environmental governance. Int Forum (05):42–47, 80
3. State Environmental Protection Administration, National Bureau of Statistics (2004) China Green National Accounting Study Report. Environ Prot 2006(09B):22–29
4. Fu J, Tang W (2010) A comparative analysis of climate change and traditional environmental problems. Chin J Rural Well-being Technol 2010:77–79
5. Xu F (2013) The role of the United Nations in global climate governance. Dissertation, Nankai University
6. UNEP Open Data Portal. https://open.unep.org/. Accessed 19 Dec 2023
7. United Nations (2022) Programme performance report of the United Nations for 2022: report of the Secretary-General. United Nations, New York
8. United Nations Environment Programme (2023) Adaptation gap report 2023: underfinanced. Underprepared. Inadequate investment and planning on climate adaptation leaves world exposed. UNEP, Nairobi
9. Ibid
10. United Nations Environment Programme (2023) Environment fund status report. October 2023. UNEP, Nairobi
11. Chenxia S (2014) The United Nations and Global Climate Change Governance: Issues and Responses. Socialism Studies 03:161–167
12. The World Meteorological Organization (WMO) is a specialized agency of the United Nations and is the UN system's authoritative voice on the state and behavior of the Earth's atmosphere, its interaction with the oceans, the climate it produces, and the resulting distribution of water resources
13. The United Nations Environment Programme (UNEP) is the leading global environmental authority within the UN system. It sets the global environmental agenda, promotes the coherent implementation of the environmental dimension of sustainable development, and serves as an authoritative advocate for the global environment

14. Kejun J (2023) IPCC AR6: Long-term Mitigation Pathways. Adv Clim Chang Res 19(02):133–138
15. Tang W, Wu L (2023) Interpretation of IPCC AR6: new perspectives in climate governance policy and implications for China. Adv Clim Change Res 19(02):151–159
16. Qihui G, Yuanyuan Q, Meicong L et al (2023) Interpretation of the IPCC AR6 Synthesis Report and Recommendations for China. Environ Prot 51(Z2):82–84
17. Zhao X, Huang M, Pang B (2022) IPCC Releases Climate Change 2021: The Physical Science Basis – Summary for All. China Meteorological News, 11 Nov 2022, p 001
18. Xin Y, Xu Y (2022) IPCC: The next few years are key to controlling global warming. China Science Daily, 7 Apr 2022, p 001
19. Hao R (2022) A study on the role, dilemma, and countermeasures of the IPCC from the perspective of epistemic communities. Dissertation, Jilin University
20. The World Wide Fund for Nature (WWF) is one of the world's largest and most experienced independent conservation organizations, with a global network active in over 100 countries. Since its founding in 1961, WWF has been dedicated to conservation efforts.
21. Schleussner C, Trisos C, Mukherji A (2023) A system-transitions report for the next IPCC assessment cycle. Nature 624(7990):42
22. Wang Z, Yuan J, Pang B, et al (2022) Key conclusions and insights from IPCC AR6 Climate Change 2022: Mitigation of Climate Change. Adv Clim Chang Res. http://kns.cnki.net/kcms/detail/11.5368.P.20220818.1738.002.html.
23. International Energy Agency (1994) The history of the IEA 1974–1994, IEA the first 20 years, vol I, origins and structure. IEA, Paris
24. International Energy Agency. The history of the IEA, vol IV, IEA, Paris
25. International Energy Agency (1994) The history of the IEA 1974–1994, IEA the first 20 years, vol III, principal documents. IEA, Paris
26. Florini A (2011) The International Energy Agency in Global Energy Governance. Global Pol 2:40–50
27. Van de Graaf T (2014) The International Energy Agency. In: Handbook of governance and security. Edward Elgar Publishing, Cheltenham, pp 489–503
28. Van de Graaf T, Lesage D (2009) The International Energy Agency after 35 years: Reform needs and institutional adaptability. Rev Int Organ 4:293–317
29. International Energy Agency. Shared goals. IEA, Paris
30. H, Hua (2005) International Energy Agency. World Petroleum Industry 12(3):66–69
31. Miller RG (2011) Future oil supply: The changing stance of the International Energy Agency. Energy Policy 39(3):1569–1574
32. Ma Y (2015) On the Relationship between China and the International Energy Agency. Contemp Int Relat 10:39–47
33. United Nations Environment Programme. "About UNEP." https://www.unep.org/zh-hans/guanyulianheguohuanjingshu/zijinhehezuohuoba. Accessed 11 Nov 2024
34. "The Carbon Footprint of Food Waste." Guangming Online, 11 Sept. 2021. https://m.gmw.cn/baijia/2021-09/11/1302569458.html
35. Xu X, Sharma P, Shu S et al (2021) Global greenhouse gas emissions from animal-based foods are twice those of plant-based foods. Nat Food 2:724–732
36. United Nations. "Sustainable Development Goal 12." Department of Economic and Social Affairs. https://www.un.org/sustainabledevelopment/sustainable-consumption-production/
37. United Nations Environment Programme (2021) Food waste index report 2021. UNEP, Nairobi, p 41
38. Ibid. (p. 51)
39. Ibid. (p. 34)
40. Ibid. (p. 55)
41. Ibid. (p. 56)
42. Ibid. (p. 78)
43. Ibid. (p. 79)
44. Ibid. (p. 81)

45. Interview with Andrew Steer on Global Environmental Governance. Sina Finance (ESG Channel), 6 June 2022. http://finance.sina.com.cn/esg/ep/2022-06-06/doc-imizmscu5169255.shtml

Chapter 2
The United Kingdom: From "Europe's Polluter" to a Global Climate Leader

In the grand narrative of global climate governance, the evolution of national roles often reflects the broader transformations in the international political economy and the shifting priorities of domestic governance. The United Kingdom—birthplace of the Industrial Revolution and one of the archetypal advanced economies—presents a paradigmatic case of policy transformation in climate governance. Over the past three decades, Britain has journeyed from reluctant participant to active global leader, crafting a model trajectory from resistance to reform to leadership. This chapter investigates the evolution of the United Kingdom's climate policy and governance framework, tracing its metamorphosis from a "troublesome member" within the European Community/European Union to an assertive and globally ambitious climate leader. It further assesses the continuity and innovations of British climate strategy in the post-Brexit era. Through this case study, we gain insights into how a sovereign state balances domestic politics, economic interests, international image, and global responsibility in shaping a distinctive climate governance pathway.

2.1 Britain Within the EU Framework: From Opposition and Acceptance to Leadership (1973–2016)

Britain's four-decade membership in the European Union left an indelible mark on its environmental and climate policies. During this period, the British approach to climate governance underwent a remarkable transformation—from skepticism and resistance to gradual acceptance, and finally to leadership within the European framework. This evolution mirrored both domestic political change and the growing institutional maturity of the EU's environmental governance architecture, which in turn exerted a reverse influence upon its member states.

X. Zou, *Thirty Years of Global Climate Governance*, Energy, Climate Change & Sustainability in the Global South, https://doi.org/10.1007/978-981-95-8457-4_2

2.1.1 The Phase of Opposition (1973–1990): Sovereignty and Cost as Core Concerns

Under successive Conservative governments, particularly during the premiership of Margaret Thatcher, Britain's attitude toward the European Community's environmental initiatives was largely antagonistic. Two core concerns underpinned this resistance: sovereignty and cost.

First, sovereignty loomed large. Historically, the United Kingdom was highly vigilant against any supranational authority perceived to erode national policy autonomy. A telling example came in 1989, when the British government refused to participate in the Hague Atmospheric Summit—an early European initiative to address climate change—and even pressured the EC to withdraw collectively. London's official rationale was explicit: "The United Kingdom cannot accept the emergence of a supranational authority that might substitute for national sovereignty in environmental matters." [1]. This stance drew a clear red line delineating Britain's climate policy autonomy.

Second, cost considerations were pivotal. The British government perceived the EC's pursuit of standardized environmental benchmarks as economically burdensome, contradicting the cost–benefit logic that underlay its own regulatory philosophy [2]. At that time, climate change had entered the scientific agenda but had not yet captured public attention or political salience. Consequently, British environmental policy favored voluntary and consultative approaches, with minimal legislative compulsion.

Britain rarely took the initiative to implement EC directives and maintained limited enthusiasm for European environmental cooperation. Its cross-border pollution problems—such as acid rain affecting Scandinavia—earned it the epithet "Europe's Polluter." [3]. Although Thatcher later became one of the first world leaders to acknowledge the threat of global warming (notably in her 1989 UN address), this rhetorical recognition did not translate into support for supranational environmental policymaking. Even her domestic "coal industry closure" policy, while incidentally reducing emissions, was driven primarily by economic and political motives rather than ecological ambition.

2.1.2 Transition and Engagement (1990–1997): Adaptation Under Internal and External Pressure

In the 1990s, British climate policy began to soften, propelled by both internal and external forces.

Domestically, environmental awareness surged. The Green Party's rising electoral popularity exerted tangible pressure on the ruling Conservatives. To shed its "polluter" image, the government sought to present itself as environmentally responsible—establishing integrated pollution control systems and supporting more stringent regulatory frameworks [4].

Internationally, Britain's participation in the 1992 Rio Earth Summit and its signing of the *United Nations Framework Convention on Climate Change (UNFCCC)* marked a decisive commitment to the global fight against climate change [5]. Yet contradictions persisted. Within the EU, London remained a vocal opponent of harmonized climate instruments, rejecting the European Commission's carbon tax proposals in 1993 and 1994. Paradoxically, at the national level, the Major government explored domestic carbon taxation (though unsuccessfully) and championed the so-called "Dash for Gas"—a rapid transition from coal to cleaner natural gas. The government also proposed an ambitious target to cut CO_2 emissions by 20% from 1990 levels by 2010 (a goal ultimately unmet).

While these efforts achieved short-term emission reductions, they deepened the country's dependence on imported gas, introducing new dimensions of energy insecurity. Nonetheless, the "Dash for Gas" became a milestone in Britain's decarbonization trajectory, illustrating how political pragmatism and environmental progress could temporarily converge.

The watershed came in 1997, when Tony Blair's Labour Party assumed power. With its pro-European orientation and commitment to integrating environmental and socio-economic goals, Labour laid the foundations for a new era in Britain's climate policy.

2.1.3 The Leadership Phase (1997–2016): from Rule-Taker to Rule-Maker

Under the Labour government, Britain's stance on EU climate policy underwent a profound transformation. Once an obstructionist, it now emerged as a proactive and often pioneering participant. Climate change attained unprecedented prominence in domestic politics and became embedded across policy sectors under the banner of "sustainable development."

This period saw Britain achieving several landmark breakthroughs that positioned it as a global leader:

International Commitments and Leadership—The Blair government played an instrumental role in the negotiation of the *Kyoto Protocol*, committing to emission reductions (12.5%) exceeding the EU average. Britain became a leading advocate of multilateral climate action.

Domestic Policy Innovation –

- In **2002**, Britain established the world's first national-level *Emissions Trading Scheme (UK ETS)*, serving as a prototype for later international carbon markets.
- In **2003**, the government's *Energy White Paper* introduced the notion of a "low-carbon economy," strategically linking economic transformation with climate mitigation.

- In **2006**, the *Stern Review*—commissioned by the Treasury—demonstrated the economic rationale for early action, asserting that the costs of inaction would far exceed those of mitigation.
- In **2008**, the *Climate Change Act* was enacted, making the UK the first country in the world to set legally binding long-term emission reduction targets—an 80% cut by 2050 from 1990 levels—and to establish the independent *Committee on Climate Change (CCC)* to ensure scientific and non-partisan policymaking [6].

Diplomatic Expansion—Britain leveraged its diplomatic networks to elevate climate security on the global agenda. Its efforts culminated in the UN Security Council's first-ever debate linking climate change and international peace—consolidating Britain's status as both a rule-maker and an intellectual leader in global climate governance.

Through these initiatives, Britain not only shed its "polluter" label but also institutionalized a comprehensive, legally grounded, and science-based governance framework that would sustain its leadership beyond the EU era.

2.2 Post-Brexit Strategic Motivations: The Logic of Sustained Ambition

The 2016 Brexit referendum triggered widespread concern that the United Kingdom might relax its climate ambition once liberated from the EU's institutional constraints. Yet, the opposite occurred. Both the May and Johnson governments reinforced, rather than retreated from, their climate commitments. This continuity—and even intensification—of ambition stemmed from a complex interplay of international and domestic motivations.

2.2.1 International Drivers

a. **Global Consensus and Green-Economy Opportunities**

Entering the 2020s, scientific consensus on climate change had reached unprecedented solidity. Successive *Intergovernmental Panel on Climate Change (IPCC)* Assessment Reports reaffirmed the anthropogenic causes of global warming, while the entry into force of the *Paris Agreement* inaugurated a new phase in global climate governance. Carbon neutrality became a shared agenda among major economies [7]. Over 127 countries—representing more than half of global GDP and greenhouse gas emissions—have pledged net-zero targets. The European Union's *Green Deal*, the Biden administration's re-entry into the *Paris Agreement* and elevation of climate diplomacy, as well as the low-carbon transitions pursued by emerging economies, together generated a worldwide "green wave."

Within this context, energy transformation and low-carbon technological innovation came to be seen as the driving engines of twenty-first-century economic growth. For post-industrial Britain, participation in this global green economy is not merely an environmental imperative but an economic necessity. Failing to maintain momentum would risk forfeiting competitiveness in sectors that are likely to define the next wave of global industrial leadership.

b. **Climate Diplomacy Under the "Global Britain" Narrative**

Post-Brexit, the British government launched the *Global Britain* strategy, aiming to project an image of a confident, outward-looking, and independent power. Faced with the need to reconstruct trade frameworks, stabilize the financial sector, and prevent erosion of international influence, climate governance offered an appealing and relatively consensual arena through which London could reassert leadership [8].

Climate diplomacy thus became a powerful instrument of foreign policy identity reconstruction. By championing a universally supported cause, the UK could rebuild partnerships with both advanced and developing economies, positioning itself as a credible and constructive actor in multilateral affairs. The hosting of the *26th Conference of the Parties (COP26)* in Glasgow provided an exceptional platform to perform this "green diplomacy," reinforcing the notion of a "Global Britain" leading by moral and technological example.

2.2.2 Domestic Drivers

a. **Economic Rationality and Green Growth**

Green Finance Leadership. London has long been one of the world's foremost centers for green finance. From the launch of its carbon trading system in 2001 to the establishment of the world's first *Green Investment Bank* in 2012, the UK built institutional and reputational advantages that have only deepened after Brexit. According to the *Global Green Finance Index (2020)*, the UK—particularly London—remains the world's top green-finance hub [9]. Preserving and expanding this dominance is critical to sustaining Britain's global financial competitiveness.

Empirical evidence demonstrates that decarbonization and economic expansion are not mutually exclusive. Between 1990 and 2018, Britain reduced its CO_2 emissions by 43% while growing its economy by 75%. Since 2000, it has achieved the fastest rate of decarbonization among G7 nations[10]. The low-carbon economy already supports more than 460,000 jobs, and projections suggest the figure could reach two million by 2030, with an annual growth rate of approximately 10% [11]. Investment in the net-zero transition is therefore not a fiscal burden but a strategic opportunity—enhancing capital formation, productivity, and technological competitiveness in the international marketplace.

b. **Political Consensus and Societal Pressure**

Cross-Party Political Agreement. Climate change represents one of the rare policy areas in Britain that enjoys durable bipartisan consensus. From Margaret Thatcher's early recognition of climate science, through Tony Blair's proactive leadership, to David Cameron's and later Conservative governments' defense of the *Climate Change Act*, both major parties have converged on viewing climate governance as a means to enhance Britain's global reputation and drive economic modernization. This shared outlook minimizes partisan conflict and provides remarkable policy stability.

Public Opinion and Civil-Society Mobilization. Public environmental consciousness has grown steadily. Groups such as the *Green Alliance*—representing around eight million citizens—have persistently lobbied the government to keep climate action at the center of the post-Brexit policy agenda [12]. Eurobarometer surveys from 2015–2017 showed a significant rise in the proportion of Britons viewing climate change as a "very serious problem." [13]. In an era of political volatility, responsiveness to environmental expectations became an essential component of governing legitimacy.

c. **Institutional and Legal Path Dependency**

The passage of the *Climate Change Act 2008* established a robust institutional architecture—most notably the *Committee on Climate Change (CCC)*—which provides authoritative advice, issues annual progress assessments, and maintains transparency. This framework ensures continuity regardless of political turnover. Any attempt to dilute climate ambition faces not only political opposition but also substantial legal obstacles. The *CCC* thus functions as a guardian of Britain's long-term climate vision.

Taken together, these international and domestic forces explain why the UK's climate strategy has intensified rather than diminished in the wake of Brexit. Reinforcing climate leadership serves multiple objectives: redefining international status, capturing the green-economy frontier, responding to social consensus, and maintaining institutional credibility. The result is not a sentimental attachment to multilateralism but a rational recalibration of national interest under new geopolitical and economic conditions. In this sense, Britain's post-Brexit climate policy is less a rupture than an evolution—an adaptive continuity in pursuit of renewed global relevance.

2.3 Post-Brexit Climate Governance: Institutional Architecture and Sectoral Implementation

To translate its ambitious climate vision into tangible results, the post-Brexit United Kingdom undertook a comprehensive reconfiguration of both institutional coordination and sectoral policy frameworks. This reform demonstrated a determination to convert the *net-zero by 2050* commitment from a rhetorical aspiration into a strategic blueprint guiding all spheres of economic and social transformation.

Strengthening High-Level Institutional Coordination

Recognizing that effective climate governance demands cross-departmental coherence, the British government moved quickly after Brexit to reinforce its top-level architecture. Cabinet Committee on Climate Change (established October 2019). Chaired directly by the Prime Minister, this committee oversees all strategies related to the net-zero transition. Its placement at the apex of executive authority ensures that climate objectives occupy the highest rung of policy priority [14]. Climate Action Strategy Committee (CAS) and Climate Action Implementation Committee (CAI, established May 2020). These two bodies form a dual-layer structure for policy-making and enforcement. The CAS formulates overarching domestic and international strategies, while the CAI—chaired by the Secretary of State for Business, Energy and Industrial Strategy—ensures operational coordination and policy delivery [15]. Together, these committees institutionalize the "whole-of-government" approach to climate governance, reducing interdepartmental frictions and ensuring consistency between high-level commitments and concrete execution.

Deep Decarbonization Across Key Sectors

The UK government recognizes that achieving net zero requires a full-scale transformation of the energy, industry, transport, building, and land-use sectors. Accordingly, it launched a series of targeted initiatives combining regulation, market incentives, and innovation support.

a. **Energy Sector: Building a Clean Power Backbone**

The decarbonization of electricity generation lies at the heart of Britain's net-zero strategy. Over the past decade, the country has made remarkable progress—achieving its lowest coal dependency since the nineteenth century—and continues to expand its clean-energy agenda through multiple initiatives:

- **Coal Phase-Out**. The UK advanced its deadline to close all remaining coal-fired power plants from 2025 to 2024, signaling an irreversible departure from the dirtiest fossil fuel [16]
- **Offshore Wind Expansion**. The government pledged to increase offshore wind capacity from 30 to 40 gigawatts by 2030 and set a new goal of 1 gigawatt in floating offshore wind by the same date [17]
- **Innovative Contract-for-Difference (CfD) Mechanism**. The CfD scheme—Britain's signature renewable support tool—has been extended to next-generation technologies, including carbon capture, utilization, and storage (CCUS), and low-carbon hydrogen production, incentivizing commercialization at scale [17]
- **Nuclear and Smart Grid Development**. Policy support for nuclear energy and the nationwide rollout of smart meters aim to create a cleaner, more flexible, and resilient energy system capable of integrating intermittent renewables.

These measures have transformed the UK into a world leader in renewable deployment, particularly offshore wind, and illustrate the synergistic relationship between industrial innovation and climate ambition.

b. Buildings Sector: Confronting the Energy Challenge of the Built Environment

The building sector accounts for roughly 19% of Britain's total greenhouse gas emissions [18]. The decarbonization challenge lies primarily in retrofitting millions of existing structures while improving standards for new construction.

- **Heat and Buildings Strategy**. The strategy mandates that new homes produce 75–80% less carbon emissions compared to current standards, advancing a long-term shift toward energy-efficient construction.
- **Financial Support Schemes**. The government launched a £3 billion *Net Zero Building Package*, including a £2 billion *Green Homes Grant* for household energy efficiency retrofits and £1 billion for decarbonizing public-sector estates such as schools and hospitals [19]
- **Innovation Pilots**. Several funds support research, demonstration, and deployment of advanced insulation materials, low-carbon heating systems, and heat-pump technologies.

While implementation has encountered administrative hurdles, these initiatives underscore the state's intention to embed climate goals in the daily lives of citizens and institutions alike.

c. Transport Sector: Transforming the Largest Emission Source

Transport remains Britain's largest emitting sector, responsible for approximately 28% of emissions in 2018 [20]. The government's core objective is to accelerate the transition to zero-emission mobility.

- **Ban on Petrol and Diesel Vehicles**. From 2030, the sale of new petrol and diesel cars and vans will cease; by 2035, all new vehicles must be zero-emission [21]
- **Infrastructure and Incentives**. £1.3 billion is allocated to expand the charging network, £582 million for consumer purchase grants, and £500 million for battery innovation and manufacturing capacity [17]
- **Sustainable Mobility Promotion**. Dedicated funds support cycling infrastructure, zero-emission public transport fleets, and urban mobility projects designed to reduce congestion and improve air quality.

This multi-tiered strategy aims not only at emissions reduction but also at reshaping Britain's industrial base—anchoring new value chains in electric vehicles, batteries, and green logistics.

d. Industrial Sector: Balancing Competitiveness and Decarbonization

Decarbonizing industry presents perhaps the greatest economic challenge: cutting emissions without undermining competitiveness.

- **Cost Compensation Schemes**. Energy-intensive industries such as steel and chemicals receive government support to offset electricity price differentials caused by carbon policies.

- **Innovation Investment**. The *Industrial Decarbonisation Strategy* mobilizes over £1 billion for low-carbon hydrogen, waste-heat recovery, and carbon capture clusters. A £2 billion portfolio of demonstration projects underpins the development of low-carbon industrial hubs, notably in Teesside and the Humber.

This approach balances the political imperative of maintaining employment and industrial strength with the environmental necessity of long-term decarbonization.

e. **Natural Resources and Land Use: Harnessing Nature-Based Solutions**

Britain's net-zero strategy also emphasizes the climate potential of land and natural resources.

- **Agriculture**. The *Environmental Land Management Scheme (ELMS)* pays farmers to deliver public goods such as carbon sequestration and biodiversity restoration, aligning agricultural subsidies with ecological outcomes.
- **Forestry**. The government committed to planting 30,000 hectares of trees annually by 2025 and introduced the *Woodland Carbon Guarantee* to provide stable revenues for forest-based carbon sinks.
- **Waste Management**. The *Resources and Waste Strategy* includes deposit-return schemes, expanded producer responsibility for packaging, and taxation on non-recycled plastic content.
- **Fluorinated Gases**. The UK adheres to the *Kigali Amendment* of the *Montreal Protocol*, phasing down hydrofluorocarbons (HFCs) and promoting leadership in sustainable cooling technologies.

Together, these measures form a coherent and multidimensional "policy mosaic." By integrating industrial, social, and environmental instruments, the UK seeks to operationalize net zero not as a distant aspiration but as a comprehensive reconfiguration of national development.

2.4 Evaluating Outcomes and Anticipating Future Challenges

After decades of evolving climate policy—particularly the intensification following Brexit—Britain's results have been impressive yet not without limitations. A critical assessment reveals both substantial progress and emerging vulnerabilities.

2.4.1 Tangible Environmental and Economic Achievements

From 1990 to 2021, Britain achieved nearly a 50% reduction in territorial greenhouse gas emissions—from approximately 805 million tonnes CO_2-equivalent to 406 million tonnes [22]. This decoupling of economic growth from emissions demonstrates the feasibility of sustainable prosperity. The most significant reductions occurred in the energy-supply sector, followed by industry [23]. The key driver was a fundamental shift in the electricity fuel mix: coal's share fell from 65.3% in 1990 to just 2.9% in 2021, while low-carbon sources—gas, nuclear, and renewables—rose from 22.2% to 50.6% [17, 24]. The low-carbon and renewable energy industries now function as engines of post-pandemic recovery. The government's *Green Industrial Revolution* plan emphasizes that by 2050, twelve priority low-carbon sectors could contribute £27 billion in domestic output and £26 billion in exports [25, 26]

2.4.2 Persistent Structural Challenges

Despite ambitious targets, gaps remain between commitment and delivery. Programs such as the *Green Homes Grant* suffered from administrative bottlenecks and low uptake. Achieving net zero will require hundreds of billions in investment—necessitating stable long-term mechanisms for mobilizing both public and private capital [27].

The 2022 Ukraine crisis exposed vulnerabilities in the UK's energy system. To ensure supply security, the government revisited North Sea oil and gas exploration and even approved a new coal mine—decisions that undermine climate credibility [28]. Balancing short-term energy needs with long-term decarbonization remains a delicate act. Britain's independent *UK Emissions Trading Scheme (UK ETS)* replaced its participation in the EU system. Ensuring that the UK carbon price remains competitive and that the market avoids "carbon leakage" poses complex regulatory challenges [29]. Moreover, the EU's *Carbon Border Adjustment Mechanism (CBAM)* will directly affect UK exporters, compelling policy coordination to avoid trade frictions.

Hosting COP26 symbolized the zenith of Britain's climate diplomacy, culminating in the *Glasgow Climate Pact*. Yet the summit also exposed limitations—particularly tensions with developing nations over climate finance [30]. Maintaining leadership without EU backing and amid domestic political volatility will test Britain's diplomatic endurance. Having exhausted the "low-hanging fruit" of coal phase-out and electricity reform, future emissions cuts will require breakthroughs in hard-to-abate sectors—steel, cement, aviation, and agriculture—where technologies such as green hydrogen and CCUS remain nascent and costly. Overreliance on unproven technologies could delay genuine systemic change.

2.5 Overall Assessment and Prospects

Britain's multi-decade trajectory in climate governance—particularly the deepened commitments of the post-Brexit period—constitutes one of the most sophisticated national experiments in reconciling environmental ambition with economic pragmatism. Its record demonstrates that decarbonization, when anchored in institutional continuity and market innovation, can coexist with sustained growth. Yet the British model also exposes structural contradictions that will shape its next phase of policy evolution.

2.5.1 Substantial Achievements in Environmental and Economic Dimensions

Data from the Department for Business, Energy and Industrial Strategy (BEIS) clearly illustrates that between 1990 and 2021, UK territorial greenhouse gas emissions fell by nearly half, even as GDP expanded substantially [17, 22]. The decoupling of economic performance from carbon intensity—long considered elusive—has now been empirically demonstrated. Britain's achievement in this regard surpasses that of most industrialized economies, validating the argument that economic modernization and ecological transition can be mutually reinforcing rather than mutually exclusive.

The reconfiguration of Britain's energy system is perhaps its most striking success. By 2021, coal's share in power generation had declined to less than 3%, replaced by a diversified portfolio of renewables, nuclear, and natural gas [18, 23, 24]. The transition was facilitated by a combination of carbon pricing, renewable subsidies, and technological learning effects—illustrating the efficacy of policy-induced innovation.

The post-pandemic policy framework branded as "Build Back Better, Greener" recast decarbonization as a national growth strategy. Offshore wind, green finance, and energy-efficient construction have become key pillars of job creation and regional revitalization. The *Green Industrial Revolution* strategy envisions 270,000 new green jobs by the mid-2020s and potentially millions by 2050 [25, 26]. In essence, Britain's low-carbon industries have evolved from environmental experiments into engines of competitiveness, demonstrating that "green" can be both an ethical imperative and an industrial advantage.

2.5.2 Enduring Challenges and Systemic Constraints

Despite remarkable progress, Britain's climate governance faces five critical tensions that will determine the sustainability of its achievements.

Ambitious targets demand consistent execution and financial realism. The failure of the *Green Homes Grant* revealed administrative inefficiencies and a lack of private-sector mobilization. Meeting the net-zero goal by 2050 requires an estimated investment of £50–70 billion annually through mid-century. Although the government has pledged substantial support, bridging this gap will depend on innovative financial instruments—green bonds, blended finance, and pension-fund mobilization—to channel long-term capital into decarbonization projects [27].

The 2022 energy crisis, exacerbated by the war in Ukraine, starkly highlighted the trade-off between short-term energy security and long-term climate commitments. The government's approval of a new coal mine in Cumbria and expanded North Sea oil exploration were politically expedient yet environmentally regressive steps [28]. This dilemma underscores the fragility of climate ambition in the face of geopolitical shocks. A credible pathway forward will require a coherent energy security doctrine that integrates renewable diversification, grid resilience, and storage capacity.

Brexit has afforded Britain greater policy autonomy but also introduced systemic friction. The establishment of the independent *UK Emissions Trading Scheme (UK ETS)* raises regulatory challenges in maintaining price alignment with the EU ETS to prevent carbon leakage [29]. Furthermore, the EU's *Carbon Border Adjustment Mechanism (CBAM)* risks penalizing UK exporters, effectively forcing Britain to synchronize its carbon pricing and verification systems to remain competitive. Rather than freeing Britain from Brussels' influence, Brexit has paradoxically tied its climate governance to the EU's evolving regulatory orbit.

COP26 in Glasgow symbolized both the zenith and the fragility of Britain's climate diplomacy. While the *Glasgow Climate Pact* reaffirmed the 1.5 °C goal, negotiations revealed persistent divides between developed and developing nations—particularly over climate finance and loss-and-damage mechanisms [30]. Without the collective bargaining power of the EU, the UK's capacity to influence global negotiations depends on sustained credibility and generous financial contributions. Political instability—exemplified by frequent changes in prime ministers—further undermines continuity in climate diplomacy.

With the "easy wins" of coal phase-out and electricity decarbonization largely achieved, the remaining emissions cuts will come from hard-to-abate sectors: heavy industry, aviation, shipping, and agriculture. The technologies required—green hydrogen, sustainable aviation fuels, and large-scale CCUS—are either immature or prohibitively expensive. Britain's reliance on unproven technologies risks deferring meaningful transformation. Policymakers must strike a balance between innovation optimism and pragmatic diversification to avoid technological lock-in.

2.6 The Political Economy of British Climate Leadership

Britain's three-decade odyssey from "Europe's Polluter" to global climate leader epitomizes the adaptive evolution of a sovereign state navigating the complex interplay between domestic politics, economic imperatives, and international identity.

The transformation was neither linear nor purely environmental; it was driven by the politics of legitimacy, competitiveness, and strategic positioning within the global order.

2.6.1 The Foundations of Success

Three pillars underpin the relative success of Britain's climate governance: First, Cross-Party Consensus and Legal Rigidity. The rare bipartisan agreement on climate policy, institutionalized through the *Climate Change Act* and safeguarded by the *Committee on Climate Change (CCC)*, has insulated climate ambition from electoral cycles. Legal obligations transformed political commitment into administrative discipline; Second, Integration of Climate and Economic Strategy. By aligning decarbonization with industrial renewal and financial innovation, Britain reframed climate policy from moral burden to economic opportunity. The rise of London as the global hub of green finance and the expansion of offshore wind clusters along the North Sea coast are tangible outcomes of this integration; Third, Diplomatic Instrumentalization of Climate Leadership**.** Post-Brexit, the UK leveraged climate diplomacy to compensate for diminished geopolitical influence, embedding environmental leadership within the broader narrative of "Global Britain." Through COP26 and related initiatives, it sought to project an image of responsible sovereignty—a state that leads not through coercive power but through normative and technological example.

2.6.2 The Lessons and Warnings

Yet Britain's story also offers cautionary lessons. Climate leadership, once achieved, must be constantly reaffirmed through consistent implementation, credible financing, and social legitimacy. The path ahead demands not only technological innovation but also distributive justice—ensuring that the burdens and benefits of transition are equitably shared across regions and classes. The challenge is no longer announcing ambition but sustaining transformation. In the coming decades, Britain's ability to navigate energy security dilemmas, social transitions, and international coordination will test the resilience of its climate governance model. If it succeeds, the UK will reaffirm that national sovereignty and global responsibility can coexist within a cooperative climate order. If it falters, it risks revealing the fragility of climate leadership in an age of geopolitical fragmentation.

2.6.3 Implications for Global Climate Governance

From a global governance perspective, Britain's experience underscores a critical insight: effective national climate action emerges from the convergence of *environmental integrity*, *economic competitiveness*, and *social inclusiveness*. Climate ambition must be embedded in a durable political economy framework rather than treated as an isolated policy domain. In the post-Brexit world, the relationship between the UK and the EU will oscillate between cooperation and competition in climate governance. Their ability to align carbon markets, coordinate green industrial policies, and jointly influence global norms will shape the future architecture of international climate order. The British case thus provides not merely a national narrative but a microcosm of the twenty-first-century struggle to reconcile sovereignty with planetary stewardship. Ultimately, Britain's climate journey remains unfinished. Its successes offer inspiration; its challenges offer warning. In the grander chronicle of global climate governance—from Rio to Glasgow and beyond—the United Kingdom stands as both protagonist and mirror, reflecting the possibilities and paradoxes of humanity's pursuit of a sustainable civilization.

References

1. Paul G (2007) Harris, Europe and global climate change: polities. Edward Elgar Publishing, Foreign Policy and Regional Cooperation, p 89
2. Weale A (1999) European environmental policy by stealth: the dysfunctionality of functionalism? Eviron Plann C Gov Policy 17(1):37–51
3. Wang Y (2019) The transformation of 'Europe's polluter': conflicts and convergence in UK–EU air pollution governance. Dongwu Acad J 4:56
4. Rayner T, Jordan A (2017) The United Kingdom: a record of leadership under threat. In: Wurzel R, Connelly J, Liefferink D (eds) The European Union in international climate change politics: still taking a lead? Routledge, p 175
5. Ibid., p.176
6. Li W (2010) The Climate Change Act and the evolution of the UK's energy and climate policy. Int Outlook 2:45–48
7. Yuan J, Zhang Y (2022) "Climate science and carbon neutrality", China population. Resour Environ 32(9):48
8. Qu B, Wang S (2021) The 'global Britain' strategy in the post-Brexit era. Mod Int Relat 1:45
9. Hu, S., op. cit., p.55
10. Ibid., p.50
11. Ibid., p.51
12. Silverwood J, Moulton J (2018) On the agenda? The multiple streams of Brexit-era UK climate policy. Marmara J Eur Stud 26(1):85
13. Special Eurobarometer 459: United Kingdom, March 2017, p.15
14. Prime Minister's Office, PM to Chair New Cabinet Committee on Climate Change, 17 October 2019
15. HM Treasury, Net Zero Review Launched to Support UK's World-Leading Climate Commitment, 2 November 2019
16. BEIS, New Plans to Make UK World Leader in Green Energy, 6 October 2020
17. Ibid

18. BEIS, Final UK Greenhouse Gas Emissions National Statistics: 1990–2018, 30 July 2020
19. HM Treasury, Budget 2020: What You Need to Know, 11 March 2020
20. BEIS, Final UK Greenhouse Gas Emissions National Statistics, 2 February 2021
21. BEIS, The Ten Point Plan for a Green Industrial Revolution, November 2020
22. BEIS, Provisional UK Greenhouse Gas Emissions National Statistics 2021, 2022
23. BEIS, Provisional Greenhouse Gas Inventory 1990–2019, 26 March 2020
24. BEIS, UK Energy Statistics, 2022 & Historical Data Series
25. Hu, S., op. cit., p.52
26. Ibid., p.53
27. Committee on Climate Change, 2022 Progress Report to Parliament, June 2022
28. The Guardian, "UK Approves First New Coal Mine in Decades, Sparking Climate Backlash," December 2022
29. The UK ETS after Brexit: Challenges and Opportunities, Climate Strategies Report, 2021
30. BBC News, "COP26: What Was Agreed at the Glasgow Climate Conference?", November 2021

Chapter 3
The European Union and Climate Governance: Evolution of Leaderships the Systems

The European Union (EU), as one of the largest economies in the world, has long been at the forefront of global climate governance. Since the 1990s, the EU has actively shaped and driven international climate policies, progressively developing a multi-layered governance system characterized by a combination of policy measures and market-based instruments. The EU's commitment to climate action is evidenced through pivotal international agreements such as the Kyoto Protocol and the European Green Deal [1]. This chapter focuses on the evolution of the EU's climate governance framework, exploring its policy development, leadership role in global negotiations, and the impact of its member states, particularly France and Germany, in shaping the EU's climate trajectory [2].

3.1 Evolution of the EU's Climate Governance System

The development of the EU's climate governance can be divided into three primary stages: the early formulation phase, the developmental phase, and the phase of maturity. Each stage represents a critical period in which the EU consolidated its institutional architecture, deepened integration between policy sectors, and progressively expanded its global influence [3].

3.1.1 The Early Formulation Phase (1980s—Early 1990s)

The EU's climate governance journey began in the mid-1980s, when the European Community (EC) first acknowledged climate change as a pressing global issue. During this period, the EC sought to align its energy policy reforms with the growing scientific consensus on global warming [4]. The 1990 European Council resolution

X. Zou, *Thirty Years of Global Climate Governance*, Energy, Climate Change & Sustainability in the Global South, https://doi.org/10.1007/978-981-95-8457-4_3

calling for "European action on climate change" marked a symbolic starting point in the EU's institutional engagement with the issue [5]. In this early phase, the European Commission emerged as a proactive advocate for climate policy coordination across member states, focusing primarily on energy innovation and efficiency [6]. These early actions coincided with international efforts under the United Nations Framework Convention on Climate Change (UNFCCC), preparing the EU for deeper engagement in the subsequent Kyoto phase [7].

3.1.2 The Developmental Phase (1990s—Early 2000s)

The 1990s were a defining decade for the EU's transformation from environmental policy coordination to comprehensive climate governance. Following the signing of the UNFCCC in 1992, the EU became an active participant in shaping global climate negotiations [8]. The Kyoto Protocol of 1997 represented a watershed moment: the EU not only committed to reducing its collective greenhouse gas emissions by 8% below 1990 levels by 2012 but also played a decisive role in drafting the Protocol's core provisions [9]. In addition to external commitments, the EU began integrating climate considerations into domestic policy domains such as transport, taxation, and industrial regulation [10]. These actions reflected a growing recognition that effective climate governance required cross-sectoral coordination rather than isolated measures.

3.1.3 The Phase of Maturity (2000s—Present)

With the entry into force of the Kyoto Protocol in 2005, the EU's climate governance entered a phase of maturity. The establishment of the EU Emissions Trading System (EU ETS) in 2005 marked the launch of the world's largest carbon market, representing a paradigm shift from voluntary cooperation to legally binding mechanisms [11]. Building on Kyoto, the EU introduced the "20/20/20" climate and energy package in 2008, setting three integrated targets for 2020: a 20% reduction in greenhouse gas emissions, a 20% increase in renewable energy, and a 20% improvement in energy efficiency relative to 1990 levels [12]. These targets embodied the EU's unique approach—linking environmental ambition to economic modernization. In 2019, the European Green Deal reaffirmed this trajectory, establishing the objective of making Europe the first climate-neutral continent by 2050 [13]. This comprehensive policy package integrated energy, industry, agriculture, and mobility under a single vision of sustainable growth, consolidating the EU's leadership role in the global transition to low-carbon development [1].

3.2 EU Climate Governance Goals and Policy Evolution

The European Union (EU) has long viewed climate governance not only as a responsibility but as a strategic opportunity to shape its future economy and global influence. The evolution of the EU's climate governance framework is deeply intertwined with its ambitious climate goals and the policies that underpin them. This section examines the development of EU climate governance goals, starting from the early objectives set under the Kyoto Protocol, through the 20/20/20 targets, to the most recent 2050 carbon neutrality goals. It will also explore the key policy frameworks that have been integral to these objectives.

3.2.1 Climate Governance Goals: Short, Medium, and Long-Term Targets

The EU's climate governance framework has been structured around specific goals and targets, which have evolved over time. These goals can be divided into short-term, medium-term, and long-term objectives, reflecting the EU's commitment to addressing climate change through a combination of emission reductions, renewable energy adoption, and energy efficiency improvements.

a. **Short-Term Goals: Kyoto Protocol and Early EU Commitments (1990s—2012)**

The short-term goals of the EU climate governance framework were initially framed by the Kyoto Protocol, which set legally binding emission reduction targets for developed countries. The EU's commitment under Kyoto was to reduce greenhouse gas emissions by 8% below 1990 levels by 2012. The Protocol established the EU's role as a global leader in climate governance, with the EU countries agreeing to make reductions in line with their national circumstances. This commitment was crucial to demonstrate the EU's commitment to tackling climate change, especially in the face of early skepticism about the effectiveness of international climate agreements [1]. In addition to these international targets, the EU also adopted several internal policies to align with its Kyoto targets. Among the most notable of these was the EU Emissions Trading System (EU ETS), which was launched in 2005. The EU ETS became the world's largest carbon market and played a pivotal role in achieving emissions reductions across the member states [2].

b. **Medium-Term Goals: The 20/20/20 Targets (2010s—2020)**

By 2007, in response to growing scientific evidence on climate change and the emerging consensus on the need for more ambitious climate action, the EU adopted its "20/20/20" climate and energy package. This package set three binding targets for 2020:

- A 20% reduction in greenhouse gas emissions compared to 1990 levels.

- A 20% share of renewable energy in total energy consumption.
- A 20% improvement in energy efficiency compared to projected levels [3].

These medium-term goals marked a significant expansion of the EU's climate governance agenda, moving beyond Kyoto's emphasis on emissions reductions to include energy efficiency and renewable energy as core pillars of climate action. The 20/20/20 targets were embedded in the EU's broader economic and energy policy, with the EU pushing for the development of a green economy driven by clean technologies and sustainable practices [4]. Despite some challenges in meeting these targets, particularly with respect to the energy efficiency and renewable energy goals, the 20/20/20 package was a key milestone in the EU's ongoing efforts to decarbonize its economy. It also helped to position the EU as a leader in global climate negotiations, particularly at the Copenhagen COP15 summit in 2009 [5].

c. **Long-Term Goals: Carbon Neutrality by 2050**

Looking further ahead, the EU has set its sights on a more ambitious long-term goal: achieving carbon neutrality by 2050. This goal, which was formally enshrined in the European Climate Law in 2021, commits the EU to reducing its greenhouse gas emissions to net-zero by mid-century, in line with the goals of the Paris Agreement [6]. To achieve this, the EU has established a range of policies aimed at decarbonizing key sectors of the economy, including energy, industry, transport, and agriculture.

The 2050 carbon neutrality target is at the heart of the European Green Deal, which aims to transform the EU's entire economy into a sustainable and low-carbon model. The Green Deal outlines a detailed roadmap to achieve carbon neutrality, including a series of measures to reduce emissions, increase the share of renewable energy, and promote the development of green technologies such as hydrogen and carbon capture and storage (CCS) [7]. The European Green Deal is seen as a strategic blueprint for achieving both environmental sustainability and economic growth through innovation and green technologies.

3.2.2 Core Policies and Legal Frameworks: Key Milestones in EU Climate Governance

To support its climate governance goals, the EU has developed a comprehensive array of policies and legal frameworks. These policies span various sectors, from energy and industry to transportation and agriculture, and have been progressively refined to ensure that the EU remains on track to meet its climate targets.

a. **The EU Emissions Trading System (EU ETS)**

The EU ETS, launched in 2005, is the centerpiece of the EU's climate policy framework. It is a cap-and-trade system that sets a limit on the total amount of greenhouse gases that can be emitted by covered sectors (such as power generation and heavy industry) and allows companies to buy and sell allowances for emissions.

The system is designed to incentivize emissions reductions by putting a price on carbon and driving investment in low-carbon technologies [8]. The EU ETS has undergone several phases, with progressively stricter caps on emissions and increased reliance on auctions rather than free allocation of allowances. By 2020, the EU ETS had expanded to include more sectors, such as aviation, and had been linked to other carbon markets in neighboring countries, further enhancing its role as a global model for carbon pricing [9].

b. **The Renewable Energy Directive and the Clean Energy Package**

In addition to the EU ETS, the EU has introduced several key legislative measures to promote renewable energy and energy efficiency. The Renewable Energy Directive, adopted in 2009 and revised in 2018, sets binding targets for renewable energy deployment in the EU. The directive mandates that by 2020, 20% of the EU's energy consumption must come from renewable sources, with specific targets for each member state based on their starting point [10]. The Clean Energy for All Europeans package, adopted in 2019, is a comprehensive set of laws aimed at accelerating the transition to a clean energy system. The package includes measures to promote energy efficiency, increase renewable energy capacity, and modernize the energy market, including the creation of a European electricity market and the implementation of the Clean Energy Transition [11].

c. **The European Green Deal and the European Climate Law**

The European Green Deal, presented in 2019, is the EU's most ambitious climate action plan to date. It aims to make Europe the world's first climate-neutral continent by 2050, in line with the EU's long-term climate goals. The Green Deal encompasses a wide range of policies, including energy transition, sustainable agriculture, circular economy, and pollution reduction [12]. To ensure that the Green Deal's goals are achieved, the EU passed the European Climate Law in 2021, which legally commits the EU to achieving net-zero emissions by 2050. The law establishes a framework for the EU to meet its climate goals and includes provisions for regular reviews and updates to ensure that progress is being made [13].

3.3 EU Leadership in Global Climate Governance

The European Union (EU) has consistently positioned itself as a leader in global climate governance, not only by shaping internal climate policies but also through its role in international negotiations. The EU's leadership stems from its commitment to ambitious climate targets, its market-driven policies such as the Emissions Trading System (EU ETS), and its diplomatic efforts to foster multilateral climate cooperation. This section examines the EU's leadership role in shaping global climate policy, its influence in international negotiations, and the key challenges it faces as it continues to lead global efforts to combat climate change.

3.3.1 The EU as a Global Leader in Climate Negotiations

The EU's leadership in global climate governance has been marked by its proactive stance in international climate negotiations. From the early days of the UN Framework Convention on Climate Change (UNFCCC) negotiations to the most recent Paris Agreement, the EU has played a pivotal role in pushing for ambitious global climate action.

a. **The Kyoto Protocol and EU's Pivotal Role**

One of the EU's first significant roles in global climate governance was during the negotiations of the **Kyoto Protocol** in the 1990s. The EU played an instrumental role in shaping the Protocol's outcomes, advocating for a binding emissions reduction framework for developed countries. The EU's commitment to the Protocol was strong, with the EU as a whole agreeing to reduce emissions by 8% by 2012, with individual member states committing to different targets based on their national circumstances [1]. At the same time, the EU also worked to integrate market-based mechanisms, such as the **Clean Development Mechanism (CDM)** and joint implementation, into the Protocol's framework. These mechanisms allowed countries to meet their targets by investing in emission reduction projects in other countries, making it a more flexible and cost-effective approach. The EU's leadership in these areas was critical to the Protocol's success, although it was later undermined by the lack of participation from key global players like the United States [2].

b. **Copenhagen Summit and the EU's Push for Global Ambition**

In 2009, the EU once again demonstrated its leadership in the international climate arena during the **Copenhagen Climate Summit (COP15)**. The EU's negotiating position was clear: it advocated for strong, legally binding commitments for emissions reductions in both developed and developing countries, with the goal of limiting global temperature rise to below 2 °C. However, the summit ultimately ended in a political agreement rather than a legally binding treaty, with major economies such as the United States, China, and India resisting the EU's proposed commitments [3]. Despite the outcome, the EU's efforts to maintain the global climate agenda in Copenhagen highlighted its determination to drive ambitious global climate action. The summit's failure was a significant challenge, but it did not diminish the EU's role as a leading actor in climate diplomacy.

c. **The Paris Agreement and the EU's Influence**

The **Paris Agreement** in 2015 marked a historic milestone in global climate negotiations. The EU played a critical role in shaping the agreement, advocating for a universal climate deal that would include both developed and developing countries. The agreement's main goal, to limit global temperature rise to well below 2 °C above pre-industrial levels, was something the EU had long pushed for [4]. In the run-up to Paris, the EU made a significant commitment to reducing its own emissions by 40% by 2030 compared to 1990 levels, and it also promised to increase its climate finance contributions to developing countries. This commitment helped set the stage for the global consensus that emerged at COP21. The EU's leadership in Paris reinforced its position as the most ambitious global climate actor, aligning its domestic climate policy with its international goals [5].

3.3.2 The EU's Global Climate Diplomacy: Challenges and Opportunities

While the EU has consistently been a driving force in global climate negotiations, it faces significant challenges as it strives to maintain its leadership position. These challenges include the need to reconcile internal divisions within the EU, the rise of climate skepticism in certain member states, and the growing influence of emerging economies like China and India in global climate governance.

a. **Internal Divisions and the Politics of Climate Policy**

One of the EU's key challenges in maintaining its leadership in global climate governance is the internal divisions among its member states. While some member states, particularly in Northern and Western Europe, are highly committed to ambitious climate action, others, especially those in Eastern Europe, have historically been more resistant to stringent climate policies. These divisions have sometimes slowed the EU's ability to present a united front in international negotiations [6]. For example, countries such as Poland, which rely heavily on coal for their energy needs, have expressed concerns about the economic impact of ambitious climate policies, leading to tensions within the EU on issues like carbon pricing and renewable energy targets. These internal divisions have made it difficult for the EU to maintain a consistent and unified negotiating position on global climate issues [7].

b. **The Rise of China and India in Global Climate Governance**

Another challenge facing the EU is the rise of China and India as major players in global climate governance. Both countries have significantly increased their commitments to climate action in recent years, with China positioning itself as a global leader in renewable energy development and India emphasizing the need for climate justice and equitable climate action for developing countries [8]. As China and India become more influential in global climate negotiations, the EU's traditional leadership role is being tested. However, the EU has sought to build alliances with these emerging economies by offering climate finance and technological support. This cooperation is crucial for the success of future international climate agreements, as global emissions reductions will depend on the active participation of all major emitters, including China and India [9].

c. **The EU's Green Diplomacy: Leveraging Climate Action for Global Influence**

Despite these challenges, the EU has continued to use climate diplomacy as a key tool in shaping global policy. By promoting climate action as a way to achieve both environmental sustainability and economic development, the EU has positioned itself as a leader in the green economy. The EU's **Green Diplomacy** has been particularly effective in building partnerships with developing countries, offering both financial and technical support to help them transition to low-carbon economies [10]. In this context, the EU has also used climate action to strengthen its geopolitical influence. By advancing its climate agenda, the EU has enhanced its role in shaping global trade and energy policies, particularly with regards to the development of green technologies and clean energy [11].

3.4 France and Germany in EU Climate Governance: Twin Pillars of European Climate Leadership

The European Union's leadership in global climate governance is anchored in the commitment and policy innovation of its member states. Among them, **France and Germany** stand as the twin pillars of the EU's climate leadership. Their historical trajectories, political cultures, and energy structures have profoundly shaped the EU's approach to climate policy. Both countries have been central in the formulation and implementation of EU-wide climate goals, yet their national strategies differ in terms of policy emphasis, instruments, and energy mix. This section explores France and Germany's respective contributions to the EU's climate governance framework and examines how their domestic experiences have influenced the evolution of European climate policy.

3.4.1 Germany: The Engine of the European Energy Transition

Germany's leadership in climate policy dates back to the late 1980s, when environmental issues first entered mainstream political discourse. The country's commitment was institutionalized in the 1990s through a combination of domestic legislation and active participation in international negotiations under the UNFCCC and Kyoto Protocol [14].

Germany was among the first countries to integrate environmental objectives into its broader energy policy, introducing measures to promote renewable energy and energy efficiency. The 1991 Electricity Feed-in Act, which guaranteed fixed prices for renewable energy producers, laid the foundation for the later Renewable Energy Sources Act (EEG) of 2000 [15]. These laws provided long-term investment certainty and helped create the conditions for Germany's **"Energiewende"** — an unprecedented national strategy to transition to a low-carbon economy.

The **"Energiewende"** represents Germany's most ambitious policy innovation, aimed at achieving a sustainable energy transition through the expansion of renewable energy, the phase-out of nuclear power, and the reduction of greenhouse gas emissions

[16]. Following the Fukushima disaster in 2011, Germany made a landmark decision to accelerate its nuclear phase-out, closing all nuclear plants by 2022 and committing to a carbon–neutral energy system by 2045 [17]. Under the Energiewende, Germany has become a global leader in renewable energy technology, particularly in wind and solar power. By 2020, renewables accounted for nearly 46% of Germany's electricity generation, compared to less than 10% two decades earlier [18]. This transition has, however, presented challenges, including intermittency in energy supply, rising electricity costs, and the need for grid modernization. Nevertheless, the Energiewende has become a model for sustainable energy transition worldwide and has heavily influenced EU policy directions.

Germany's leadership extends beyond domestic policy into EU-level and international diplomacy. As Europe's largest economy and a key driver of EU integration, Germany has consistently advocated for ambitious climate targets within the EU framework. It played a crucial role in negotiating the EU's 2030 Climate and Energy Framework and was a central force behind the European Green Deal [19]. Germany's diplomatic influence is amplified through the **"Petersberg Climate Dialogue"**, an annual forum launched in 2010 to promote global cooperation in the run-up to and beyond the Paris Agreement. By providing political and financial support to developing countries, Germany has also enhanced the EU's role in global climate financing [20].

3.4.2 France: From Environmental Diplomacy to Climate Legislation

France's climate governance has evolved through a combination of **"state-driven environmentalism"** and **"strategic diplomacy"**. Unlike Germany, France's energy system has long been dominated by nuclear power, which accounts for over 70% of its electricity production. This unique energy structure has allowed France to maintain relatively low-carbon electricity generation, though it has also constrained the pace of renewable energy expansion [21].

France's modern climate policy began taking shape in the early 2000s with the **National Climate Plan (2000)** and was further reinforced by the **Grenelle Environment Forum (2007–2010)**, which introduced major reforms in energy efficiency, building standards, and sustainable transportation [22].

The Paris Agreement and French Climate Diplomacy

France's most significant contribution to global climate governance came through its leadership of the **2015 Paris Climate Conference (or COP21)**. As host and diplomatic mediator, France played a decisive role in brokering the **Paris Agreement,** which established a universal framework for global climate action. French diplomacy emphasized inclusiveness, flexibility, and ambition—principles that proved crucial to securing agreement among 195 countries [23]. The success of COP21 solidified

France's reputation as a global leader in climate diplomacy. Domestically, it also spurred a wave of legislative reforms, including the **Energy Transition for Green Growth Act (2015)**, which set ambitious goals to reduce greenhouse gas emissions by 40% by 2030 and increase renewable energy to 32% of final energy consumption [24].

France's Post-Paris Climate Initiatives

Since 2015, France has continued to strengthen its climate governance through successive national strategies. The **French Climate and Resilience Law (2021)** builds upon the Paris Agreement commitments by embedding climate objectives across all sectors of government and setting legally binding carbon budgets [25]. France has also been active in promoting climate finance and international cooperation, notably through initiatives such as the **One Planet Summit,** co-hosted by President Emmanuel Macron to mobilize global investments for climate adaptation and biodiversity conservation [26]. France's leadership in international climate diplomacy has been complemented by its efforts to align economic recovery and green transition, particularly in the wake of the COVID-19 pandemic, where France introduced a €100 billion "**France Relance**" plan emphasizing green growth and low-carbon innovation [27].

3.4.3 Comparative Analysis: Complementary Strengths and Divergent Paths

While both France and Germany have been central to EU climate governance, their paths to decarbonization reveal different political, economic, and institutional logics.

- **Policy Orientation**: Germany's approach has been market-driven and decentralized, emphasizing citizen participation and innovation through the renewable energy sector. France, by contrast, has relied on a centralized, state-led model, using legislation and planning to drive its climate transition [28].
- **Energy Mix**: Germany's dependence on coal and natural gas has made its transition more complex, whereas France's reliance on nuclear energy has provided a head start in reducing emissions but also raised questions about nuclear waste and long-term sustainability [29].
- **EU Integration**: The Franco-German partnership remains the backbone of EU climate governance. Germany's technological and industrial capacity complements France's diplomatic leadership, enabling the EU to combine domestic ambition with international legitimacy [30].

Together, France and Germany have provided both the political will and policy innovation that underpin the EU's climate leadership, creating a model of "**dual leadership**" that balances economic modernization with environmental responsibility.

References

1. United Nations (1998) Kyoto Protocol to the United Nations Framework Convention on Climate Change
2. Ellerman AD, Buchner BK (2007) The European Union Emissions Trading System: Origins, allocation, and early results. Rev Environ Econ Policy 1(1):66–87
3. European Commission (2008) Climate and energy package 20/20/20. Brussels
4. Jordan A, Rayner T (2010) The evolution of climate policy in the European Union. Environ Polit 19(2):193–210
5. Gupta J, Grubb M (eds) (2000) Climate change and European leadership: a sustainable role for Europe? Springer, Dordrecht
6. European Commission (2019) The European Green Deal. Brussels
7. European Commission (2021) European climate law: regulation (EU) 2021/1119. Brussels
8. Skjærseth JB, Wettestad J (2008) EU emissions trading: Initiation, decision-making and implementation. Ashgate, Aldershot
9. European Parliament and Council (2019) Clean energy for all Europeans package. Brussels
10. European Commission (2018) Renewable energy directive. Brussels
11. European Commission (2020) Clean energy for all Europeans—the 2030 climate and energy framework. Brussels
12. Commission E (2019) A clean planet for all: a European strategic long-term vision for a prosperous, modern. Competitive and Climate-Neutral Economy, Brussels
13. European Climate Foundation (2021) Achieving net zero by 2050: the European climate law and its implications
14. Jänicke M, Wurzel RKW (2019) Germany's Energiewende and the European Energy Transition. J Environ Plan Policy Manag 21(5):567–580
15. Lauber V, Jacobsson S (2016) The politics and economics of constructing, contesting and restricting socio-political space for renewables. Environ Innov Soc Trans 18:147–163
16. Agora Energiewende (2015) Understanding the energiewende. Berlin
17. Federal Ministry for Economic Affairs and Climate Action (BMWK) (2022) Germany's energy transition. Berlin
18. International Energy Agency (IEA) (2020) Germany 2020 energy policy review. Paris
19. Oberthür S, Roche Kelly C (2008) EU leadership in international climate policy: achievements and challenges. Int Spect 43(3):35–50
20. Petersberg Climate Dialogue (2021) Advancing global climate action through cooperation. Berlin
21. Cour des Comptes (2018) The cost and future of French nuclear power. Paris
22. Ministère de la Transition Écologique (2010) Grenelle de l'environnement: bilan et perspectives. Paris
23. United Nations Framework Convention on Climate Change (UNFCCC) (2015) The Paris agreement. New York
24. French National Assembly (2015) Loi relative à la transition énergétique pour la croissance verte (Energy transition for green growth act). Paris
25. French National Assembly (2021) Loi climat et résilience (Climate and resilience law). Paris
26. Élysée Palace (2020) One planet summit: accelerating global climate action. Paris
27. Ministère de l'Économie et des Finances (2021) France relance: the recovery plan for a green and competitive economy. Paris
28. Szarka J (2010) France's sustainable development strategy: ambition, complexity and coordination. Environ Polit 19(3):403–421
29. Fouquet D (2020) The French energy transition: between nuclear dependence and renewable ambitions. Energy Policy 138:111253
30. Wurzel RKW, Connelly J, Liefferink D (eds) (2017) The European Union in international climate change politics: still taking a lead? Routledge, London

Chapter 4
The United States—A Pendulum of Power in Global Climate Governance

The trajectory of the United States in global climate governance over the past three decades presents a paradox. As the world's largest historical emitter of greenhouse gases and a scientific and technological powerhouse, its leadership is indispensable for any effective global response. Yet, its policy path has been marked not by steadfast leadership, but by profound volatility, oscillating between engagement and retreat. This chapter examines the evolution of U.S. climate policy from the administration of George H.W. Bush to that of Joe Biden, arguing that its erratic nature is not a series of random fluctuations but a reflection of deep-seated structural tensions within the American political economy. The analysis will dissect the influential factors—economic interests, partisan polarization, federalism, and cultural norms—that drive this pendulum swing, and assess the consequent challenges and prospects for U.S. climate action on the global stage.

4.1 The Erratic Trajectory of U.S. Federal Climate Policy

The U.S. approach to international climate governance has been fundamentally shaped by the priorities of its sitting president, leading to a pattern of policy reversal with each partisan transition.

4.1.1 The Bush Administration (1989–1993): Reluctant Engagement and Sovereignty Concerns

The George H.W. Bush administration approached global climate cooperation with caution. While it extended the Clean Air Act domestically and signed and ratified the

X. Zou, *Thirty Years of Global Climate Governance*, Energy, Climate Change & Sustainability in the Global South, https://doi.org/10.1007/978-981-95-8457-4_4

United Nations Framework Convention on Climate Change (UNFCCC), it explicitly rejected any weakened obligations for developing countries [1]. Despite the emerging scientific consensus from the IPCC's First Assessment Report, the administration cited scientific uncertainty and complexity as reasons to resist immediate, binding emission reduction targets. A primary concern was economic competitiveness; internal estimates suggested that cutting CO_2 emissions by 20% could cost the economy between $800 billion and $3.6 trillion [2]. Consequently, the U.S. negotiating stance deliberately weakened the binding force of the UNFCCC, favoring adaptation strategies over mandatory mitigation.

4.1.2 The Clinton Administration (1993–2001): Symbolic Internationalism and Domestic Gridlock

The Clinton administration marked a rhetorical shift, publicly acknowledging the role of human activity in climate change. It played an active role in the negotiations leading to the 1997 Kyoto Protocol, and the final agreement reflected significant compromises to secure U.S. involvement. However, domestic politics severely constrained Clinton's ambitions. The 1997 Byrd-Hagel Resolution, passed 95–0 by the Senate, declared that the U.S. should not be a signatory to any agreement that did not mandate commitments for developing countries or would seriously harm the U.S. economy [3]. Faced with this potent political warning, Clinton signed the Kyoto Protocol but never submitted it to the Senate for ratification. This period thus epitomized a schism between international activism and domestic paralysis.

4.1.3 The George W. Bush Administration (2001–2009): Outright Rejection and Unilateralism

The administration of George W. Bush represented a sharp turn towards climate denial and unilateralism. In 2001, it formally renounced the Kyoto Protocol, criticizing its exemption for major developing economies and emphasizing the potential economic costs, with models projecting GDP losses of 1–2% [4]. In its place, the administration proposed the voluntary "Clear Skies & Global Climate Change Initiatives," favoring intensity-based targets and technology partnerships over internationally binding commitments. This approach reflected a prioritization of short-term economic interests and a deep-seated aversion to multilateral constraints on U.S. sovereignty.

4.1.4 The Obama Administration (2009–2017): the Return of Multilateral Ambition

The Obama administration sought to reposition the U.S. as a climate leader. It framed climate action not just as an environmental imperative but as an economic opportunity and a matter of national security. Domestically, its efforts centered on the 2013 Climate Action Plan, which leveraged the regulatory authority of the Environmental Protection Agency (EPA) to set carbon standards for power plants and vehicles [5]. Internationally, Obama engaged in vigorous diplomacy, culminating in the U.S.'s pivotal role in securing the Paris Agreement in 2015. The Paris Agreement's flexible, "bottom-up" architecture of Nationally Determined Contributions (NDCs) was, in part, designed to circumvent the U.S.'s domestic ratification hurdles, allowing Obama to join through executive authority.

4.1.5 The Trump Administration (2017–2021: "America First" and Climate Denial

The Trump administration constituted the most dramatic reversal. It openly dismissed climate science as a "hoax" and pursued a systematic "de-climatization" agenda [6]. Key actions included announcing the withdrawal from the Paris Agreement, dismantling the Obama-era Clean Power Plan, and aggressively promoting fossil fuel production, including coal, under the banner of "energy dominance." This policy shift was ideologically driven, prioritizing the perceived interests of traditional industries and aligning with a nationalist, anti-globalist base.

4.1.6 the Biden Administration (2021-Present): the Ambitious Reset

President Biden moved swiftly to restore climate change as a central pillar of U.S. foreign and domestic policy. On his first day in office, he recommitted the U.S. to the Paris Agreement. His administration announced an ambitious new NDC target of reducing greenhouse gas emissions by 50–52% below 2005 levels by 2030 and launched a whole-of-government approach, envisioning a $2 trillion investment in clean energy and infrastructure [7]. The strategic goal was clear: to reclaim global leadership, stimulate a green economic recovery, and ensure the U.S. competes for dominance in the clean technologies of the future.

4.2 The Structural Drivers of Policy Volatility

The whiplash-inducing shifts in U.S. climate policy are not arbitrary; they are the product of powerful, interconnected forces within the American system.

4.2.1 The Primacy of Economic Interests

Economic calculations have perpetually underpinned the U.S. stance. The fear of economic disadvantage, whether from compliance costs under the Kyoto Protocol or imagined constraints from the Paris Agreement, has been a consistent refrain, particularly among Republicans. However, the economic landscape is now bifurcated. The traditional fossil fuel industry, with its deep political connections, has long been a powerful force for inaction—exemplified by its influence during the Trump administration. Conversely, a formidable coalition of new economic actors has emerged. Major technology corporations (e.g., Apple, Google), consumer-facing brands (e.g., Starbucks, PepsiCo), and Wall Street giants (e.g., BlackRock, major banks) have publicly supported climate action, set net-zero targets, and called for ambitious federal policy [8]. This corporate alignment has reduced the political cost of climate action for leaders like Biden, who frame it as an engine for job creation and technological innovation.

4.2.2 Partisan Polarization and Ideological Rigidity

Climate change has become a hyper-polarized issue in the U.S., effectively a litmus test for party affiliation. The ideological divide between the Democratic Party's environmentalism and the Republican Party's skepticism (and often denial) has solidified. This polarization is reflected in Congress, where federal climate legislation has been perpetually stymied, and in the electorate, where views on climate change are strongly predicted by partisan identity. The result is a "consistency of partisanship" where policy is almost entirely determined by which party controls the White House and Congress, preventing the long-term, stable policy signals necessary for a profound economic transition [9].

4.2.3 *The Power of Subnational Federalism: States and Cities as Counterweights*

In the face of federal inaction or regression, state and local governments have emerged as critical actors. When Trump announced the U.S. withdrawal from the Paris Agreement, a coalition of 25 states formed the "U.S. Climate Alliance," committing to uphold the agreement's objectives. Furthermore, networks like "Climate Mayors," encompassing over 400 cities, have advanced local climate action plans, renewable energy mandates, and regional carbon markets [10]. This decentralized action demonstrates a significant reservoir of climate commitment within the U.S., creating a de facto national policy floor and providing a foundation for future federal efforts. However, it also leads to a patchwork of regulations and cannot fully compensate for the absence of a coherent national strategy.

4.2.4 *Cultural and Philosophical Underpinnings*

Deeply held cultural beliefs shape the American resistance to climate policy. A tradition of libertarianism fosters suspicion of government regulation. A strong faith in market mechanisms leads to a preference for technological solutions over behavioral change or regulatory mandates. Furthermore, a streak of unilateralism fuels skepticism of binding international agreements, which are seen as threats to national sovereignty. These cultural currents, often amplified by conservative media, provide a fertile ground for political movements that oppose concerted climate action, making sustained, long-term policy commitment exceptionally difficult.

4.3 Challenges and Prospects in the Biden Era and Beyond

Despite the ambitious reset undertaken by the Biden administration, the path forward for U.S. climate governance remains fraught with profound challenges. The nation's ability to meet its commitments and exert sustained global leadership is tested by a combination of deep-seated domestic constraints and complex international dynamics.

4.3.1 *The Persistent Domestic Gridlock*

The most significant impediment to durable U.S. climate policy remains the intense partisan polarization in Congress. While the Biden administration managed to pass the Inflation Reduction Act (IRA) of 2022—a landmark piece of legislation representing the nation's largest-ever investment in climate and clean energy—its passage was achieved along strict party-line votes. This underscores the continued absence of a durable, cross-partisan consensus on climate change. The failure to pass more comprehensive climate legislation, such as the Clean Electricity Performance Program, highlights the limits of executive and narrow partisan action. The prospect of a politically divided or opposition-controlled Congress effectively blocks the possibility of further major climate legislation, leaving executive actions vulnerable to legal challenges and reversal by future administrations. This legislative inertia directly constrains the scale and predictability of U.S. climate finance, both domestically and internationally, creating uncertainty for investors and global partners alike.

4.3.2 The Shadow of Political Reversibility

The Trump presidency exposed the fragility of U.S. climate commitments built primarily on executive authority. A future administration sympathetic to Trump's "America First" agenda could once again withdraw from the Paris Agreement, dismantle regulatory structures, and halt the implementation of programs under laws like the IRA. This "pendulum effect" severely damages the credibility of the United States as a reliable international partner. Allies and adversaries alike are forced to factor in the possibility of another U.S. retreat, which undermines global ambition and encourages other nations to hesitate in their own commitments. The constant threat of reversal acts as a major drag on the long-term planning and investment necessary for a clean energy transition, both within the U.S. and globally.

4.3.3 The Complexity of a Federated System

While state-level action has been a critical counterweight to federal inaction, it also presents challenges. The patchwork of state policies—from California's cap-and-trade system to Texas's wind power boom to the resistance of fossil-fuel-dependent states—creates an uneven regulatory and economic landscape. This can lead to inefficiencies and compliance complexities for businesses operating across state lines. Furthermore, the most ambitious state-level actions cannot fully replicate the impact of a unified national strategy, particularly on issues like a national carbon price or economy-wide emissions standards. The reliance on subnational actors, though vital, is a second-best solution to coherent federal leadership.

4.3.4 *Strained International Cooperation and Geopolitics*

The Biden administration's efforts to restore U.S. climate leadership occur within a vastly different geopolitical context from the Obama era, characterized by strategic competition with China. Although the two countries have managed to engage in limited climate diplomacy, recognizing their shared responsibility, this cooperation remains fragile and highly susceptible to broader tensions over trade, technology, and security. A deterioration in the bilateral relationship, as seen during the Trump years and potentially exacerbated under future U.S. administrations, could sever this critical channel of dialogue and collaboration. Given that the U.S. and China are the world's two largest emitters and economies, their inability to cooperate would represent a catastrophic failure for global climate governance, making the goals of the Paris Agreement nearly unattainable.

4.3.5 *Prospects: A Contested and Fragile Pathway*

Notwithstanding these formidable challenges, the landscape is not entirely bleak. The Biden administration's policies, particularly the IRA, have unleashed powerful market forces. The significant subsidies for clean energy manufacturing and deployment are catalyzing massive private investment, potentially "locking in" aspects of the energy transition by creating powerful new economic constituencies with a vested interest in a clean energy future [11]. The continued activism of states, cities, and corporations ensures a baseline of climate action that cannot be easily erased.

The long-term prospect, therefore, is one of a contested and uneven transition. Progress will likely continue, driven by subnational actors, technological cost reductions, and corporate strategy, but its pace and sufficiency will be heavily influenced by the volatile rhythm of national politics. The United States may continue to be an indispensable yet unreliable partner in global climate governance, its contributions fluctuating with the electoral cycle rather than being guided by a stable, long-term national strategy.

4.4 The American Climate Dilemma in a Warming World

The journey of the United States through three decades of global climate governance reveals a story of immense power coupled with profound internal contradiction. Its path from the cautious engagement of George H.W. Bush to the ambitious rhetoric of Joe Biden, punctuated by periods of outright denial and withdrawal, is more than a sequence of policy changes; it is a reflection of a nation grappling with its identity in a globalized, interdependent world confronting a collective threat.

The volatility of U.S. climate policy is not an anomaly but a direct outcome of its unique political anatomy. It is a product of a highly polarized partisan system where climate change has been transformed from a scientific issue into a cultural and ideological battleground. It is driven by a persistent economic tension between the powerful, entrenched interests of a fossil-fuel-based past and the rising, yet still consolidating, forces of a clean-energy future. It is shaped by a federalist structure that allows for innovative subnational action but struggles to produce a coherent, national response. And it is constrained by a political culture deeply skeptical of government intervention and multilateral commitments. This internal struggle has global consequences. The repeated failure of the U.S. to provide consistent, predictable leadership has created a vacuum in global climate politics, slowed international progress, and eroded trust among both allies and developing nations. The world watches the American pendulum swing with a mixture of hope and apprehension, aware that the trajectory of global warming will be significantly determined by the ability of the American political system to overcome its own divisions.

Looking ahead, the central challenge for the United States is to construct a climate governance framework that can withstand the volatility of its political cycles. This requires building durable institutions and policies that are less vulnerable to partisan shifts, perhaps through legislation that creates powerful economic incentives rather than easily reversed regulations. It necessitates forging a broader, more resilient political coalition that can eventually depoliticize climate action and treat it as a core national interest. Finally, it demands a clear-eyed foreign policy that can compartmentalize climate cooperation from strategic competition, particularly with China, recognizing that on this existential issue, no nation can succeed alone.

The American climate dilemma is a microcosm of a broader global challenge: the mismatch between the long-term, collective nature of the climate threat and the short-term, sovereignty-focused nature of the international system. Until the United States can resolve its own internal contradictions, it will remain the world's most pivotal and most unpredictable actor in the fight against climate change, a source of both immense potential and profound frustration for the future of the planet.

References

1. Harris PG (2007) Europe and global climate change: politics. Edward Elgar Publishing, Foreign Policy and Regional Cooperation
2. Schmidt K (1991) Industrial countries' response to global climate change. Environment and Energy Study Institute
3. Byrd-Hagel Resolution (1997) S Res 98, 105th Congress
4. Ma J (2013) US climate governance policy and its dilemmas. Am Stud 27(4):72–96. (马建英. 美国的气候治理政策及其困境[J]. 美国研究,2013,27(04):72–96+6)
5. Liu Y (2015) The development and evolution of climate governance policy under the Obama administration. Contemp World (12):72–75. (刘元玲. 美国奥巴马政府气候治理政策的发展与演变[J]. 当代世界,2015,(12):72–75)

6. Zhang T (2017) An analysis of the trajectory of U.S. climate policy under the Trump administration [in Chinese]. Peace Dev (1):51–66. (张腾军.特朗普政府的美国气候政策走向分析[J]. 和平与发展,2017(1):51–66+122)
7. Zhao X (2021) The climate agenda of the Biden administration and its impact [in Chinese]. Contemp World (5):26–33. (赵行姝. 拜登政府的气候新政及其影响 [J]. 当代世界, 2021, (05): 26–33)
8. We Mean Business Coalition (2021) Open letter to the Biden administration on climate ambition
9. Liu Q (2010) On the influence of interest groups on U.S. climate policy formulation. Int Stud (3):58–64. (刘卿. 论利益集团对美国气候政策制定的影响 [J]. 国际问题研究, 2010, (03): 58–64)
10. United States Climate Alliance (2023) Member states and progress. https://www.usclimatealliance.org/. Accessed 10 Oct 2024
11. Zhang J, Chen Y, Zhang Y (2023) Biden administration's climate policy: motivation, framework, and defects. Peace Dev 4:25–50

Chapter 5
China in Global Climate Governance: Three Decades of Evolution from Participant to Leader

5.1 The Rise of China in the Global Climate Regime

Over the past three decades, global climate governance has evolved from a peripheral environmental concern into one of the central pillars of world politics. Since the early 1990s, when the international community first began to institutionalize collective action under the *United Nations Framework Convention on Climate Change* (UNFCCC), the architecture of global environmental governance has undergone a profound transformation. Amid this transformation, few actors have had a more significant—and at times controversial—impact than the People's Republic of China. From a cautious participant in the early years of global climate diplomacy to a proactive contributor and, increasingly, a leading architect of governance norms, China's trajectory epitomizes both the opportunities and the contradictions of governing a shared global common in an age of multipolarity and interdependence.

China's evolving role in global climate governance cannot be understood merely as a linear narrative of "rise" or "responsibilities." It must instead be read as a dynamic interplay between domestic development imperatives, international systemic changes, and the country's shifting identity within the international order. Since the 1990s, two structural variables have fundamentally shaped the landscape of global climate politics: globalization and China's rise. Globalization has deepened the interdependence of economies, diffused industrial and energy-intensive production, and created a truly planetary scale of environmental risk. China's rise, in turn, has recalibrated the material and normative balance within global governance, forcing a renegotiation of the long-standing North–South divide and of the very meaning of "common but differentiated responsibilities" (CBDR). It is this mutual constitution—between systemic globalization and China's national transformation—that frames the past thirty years of global climate politics [1].

In the early 1990s, when the UN convened the Earth Summit in Rio de Janeiro, China's GDP ranked below the world's top one hundred; the country remained largely outside the dominant structures of the global economy. Its participation in

X. Zou, *Thirty Years of Global Climate Governance*, Energy, Climate Change & Sustainability in the Global South, https://doi.org/10.1007/978-981-95-8457-4_5

climate negotiations at that time was therefore informed by a developmental logic—defending economic sovereignty and prioritizing poverty alleviation over environmental obligations. Yet by the 2010s, China had become the world's second-largest economy and the largest emitter of greenhouse gases. Such a dramatic shift in its material profile transformed not only its capabilities but also international expectations of its behavior. No longer could China be viewed as a passive beneficiary of global public goods; it was now expected to be a provider of global environmental stability. The dialectic between external expectation and internal transformation lies at the heart of China's changing climate identity.

5.2 The Changing Structure of Global Climate Politics

The evolution of the global climate regime has paralleled broader shifts in international order. From the bipolar stasis of the Cold War to the fluid multipolarity of the twenty-first century, climate governance has increasingly become a venue for contesting leadership, legitimacy, and justice. The UNFCCC (1992), the Kyoto Protocol (1997), and the Paris Agreement (2015) constitute three constitutional moments of this regime, each reflecting distinct configurations of power and principle. In each phase, China's engagement both responded to and reshaped the underlying balance between developed and developing worlds.

During the 1990s, China's position was defined by defensive multilateralism: insisting on the inviolability of the CBDR principle and opposing any mandatory emissions reductions that might constrain its developmental space. This stance was not an act of obstructionism but an expression of the structural inequities embedded within the global economy. China and many developing states argued that historical responsibility for climate change rested primarily with industrialized countries whose cumulative emissions since the Industrial Revolution dwarfed those of the Global South [2]. Hence, justice demanded that mitigation obligations be differentiated according to capacity and responsibility. By the 2000s, however, the international environment had changed. The post–Cold War liberal order promoted deeper integration of emerging economies into global markets, while transnational environmental norms began to penetrate national policy-making. China's integration into the World Trade Organization (2001) symbolized not only economic globalization but also the onset of a new era in which China's developmental model was subject to unprecedented global scrutiny. This period saw the rise of "climate diplomacy" as a major foreign-policy domain, with China increasingly expected to contribute to global public goods commensurate with its status.

In the 2010s, the rise of "post-Western" multilateralism coincided with a legitimacy crisis in the traditional architecture of climate governance. The failure of the Copenhagen Conference (2009) exposed deep divisions between North and South, while the Paris Agreement (2015) sought to transcend that dichotomy through a bottom-up architecture of *Nationally Determined Contributions* (NDCs). China played a decisive role in bridging the gap between developing and developed countries, promoting a flexible yet principled compromise that enabled universal participation. In this process, China transitioned from a defensive actor to an agenda-setting power—an evolution emblematic of its broader repositioning within global governance.

5.2.1 Analytical Framework: Interests, Power, and Roles

Understanding China's transformation in climate governance requires a theoretical lens capable of integrating material, ideational, and structural dimensions. As the literature in international political theory suggests, five key variables influence a state's external behavior: the international system, national power, national reputation, decision-makers' cognition, and international expectations [3]. These variables interact to shape a state's conception of national interest and its self-ascribed role within the international community.

- National Interest

From a realist perspective, a state's behavior is primarily driven by the pursuit of its comprehensive national interest, which itself derives from the configuration of the international system and the distribution of capabilities within it. As Kenneth Waltz observed, systemic structure "acts as a constraining and shaping force" on national action, yet shifts in unit-level power distribution can alter that very structure [4]. For China, climate governance was initially framed as an extension of development strategy: protecting the right to industrialize and resisting externally imposed constraints. Over time, however, as the costs of environmental degradation became domestically salient, climate policy was reframed as integral to national modernization and global legitimacy.

- National Power and Reputation

The dual concept of power status—encompassing both material capabilities and international reputation—offers a nuanced measure of influence. As Yan Xuetong and others have argued, national power consists of resource-based capacities (economic, military, and cultural) and operational power (political capacity to mobilize and utilize these resources). Reputation, by contrast, reflects the perception of reliability and credibility in the eyes of other actors. China's growing material power after 2000 enhanced its structural leverage in climate negotiations, but reputation—trustworthiness as a responsible stakeholder—became equally crucial. Climate diplomacy thus functioned as a testing ground for China's aspiration to rule-conforming leadership rather than rule-challenging revisionism.

- National Role Conception

Holsti's concept of "national role conception" provides a bridge between systemic incentives and individual decision-making. It captures the cognitive dimension of how leaders perceive their country's appropriate place in the international order. China's evolving self-conception—from "developing country advocate" to "responsible major power" and ultimately to "normative leader"—has been a key driver of its behavioral change in climate governance. The transformation of role conception has also influenced the expectations of other states, generating a reciprocal dynamic between perception and performance. Taken together, these variables create a feedback loop between systemic structure, domestic development, and international identity. China's climate diplomacy is not a linear progression from compliance to leadership but rather a process of co-evolution between the state and the global governance regime. As the regime matures, it imposes new norms that reshape China's preferences; as China's power grows, it in turn redefines the regime's trajectory. This mutual constitution—what might be termed interactive structuralism—is the defining feature of China's participation in global climate governance over the past thirty years.

5.2.2 The Rationale for China's Engagement

At the most fundamental level, China's engagement in global climate governance stems from the interaction of three motivational logics: First, the Logic of Developmental Legitimacy. Domestic modernization remains the cornerstone of the Chinese state's political legitimacy. Participation in global environmental governance thus serves both as an external constraint and as an internal legitimizing mechanism. By demonstrating commitment to sustainable development, the government enhances its domestic authority and international image; Second, the Logic of Strategic Responsibility. As China's relative power expanded, maintaining a "responsible image" became essential to mitigating the "China threat" discourse. Active participation in climate negotiations offered a diplomatic avenue to project soft power and to showcase willingness to shoulder global obligations commensurate with material strength; Third, the Logic of Normative Leadership. Since the 2010s, Chinese leaders have increasingly articulated a global vision centered on the Community of Shared Future for Mankind and the Ecological Civilization paradigm. These frameworks position China not merely as a rule-taker but as a rule-shaper, offering what Chinese scholars call "China's solution"to global problems [5]. Climate governance thus becomes both a laboratory and a showcase for China's normative entrepreneurship.

5.3 The Evolution of China's Role in Global Climate Governance (1990–2021)

5.3.1 The Cautious Participant (1990–2000)

China's first decade of involvement in global climate governance coincided with the immediate post–Cold War transformation of the international system—from a bipolar structure to an emerging unipolar world dominated by the United States. During this period, China's GDP ranked outside the top 100, and its economic foundation remained fragile. The country was still in the early stages of industrialization, burdened with vast developmental needs and limited institutional experience in multilateral diplomacy. In this context, China's participation in the nascent climate regime was cautious, defensive, and primarily oriented toward protecting developmental sovereignty [6]. When the UNFCCC was adopted in 1992, China became one of its earliest signatories. Yet the principles that guided its participation were clear: climate change was viewed as a "low politics" issue—secondary to economic growth, poverty alleviation, and national modernization. China's negotiators played an active role in codifying the principle of "common but differentiated responsibilities" (CBDR), which institutionalized a dual structure distinguishing between the obligations of developed and developing countries [7]. Under this principle, developing countries, including China, were not bound by mandatory emissions targets, while industrialized nations were expected to take the lead in mitigation and to provide financial and technological assistance.

This normative stance reflected both pragmatic considerations and a broader worldview shaped by developmental justice. Chinese policymakers consistently argued that "emission rights are development rights," emphasizing that the historical responsibility for global warming lay with the industrialized North. As one senior Chinese negotiator later recalled, "We insisted that climate action should not come at the expense of development. Poverty eradication remains the first priority for developing countries." [8] In 1990, China established its National Climate Change Coordination Group, followed by specialized working groups focusing on impact assessment, policy coordination, and international negotiation strategies [9]. Yet technical and informational limitations constrained China's early performance. Most climate-related data came from Western sources, and China's own scientific infrastructure was still underdeveloped. The lack of technical expertise and experience in international law left Chinese delegations at a disadvantage in early negotiations.

The 1997 Kyoto Protocol marked a turning point in the institutionalization of global climate governance. However, its rigid architecture—imposing legally binding emissions targets on Annex I (developed) countries while exempting others—also revealed underlying North–South tensions. Western states, led by the United States, began to pressure China to assume greater responsibility, arguing that its rapid industrial growth and increasing emissions warranted a stronger commitment. China firmly rejected this view, warning that premature obligations would constitute a form of "developmental containment." [10] Despite its defensive posture, China fulfilled its treaty obligations conscientiously and avoided obstructionism. It continued to emphasize multilateral cooperation, capacity-building, and technology transfer as the pillars of equitable governance. The cautious attitude of the 1990s thus represented not reluctance but prudence: China sought to integrate itself into the global regime without compromising its developmental autonomy.

5.3.2 The Open and Proactive Partner (2001–2010)

The first decade of the twenty-first century witnessed China's transition from a defensive posture to a more open and proactive engagement in global climate governance. This shift coincided with major domestic transformations and external realignments: China's accession to the World Trade Organization (WTO) in 2001, rapid industrial expansion, and growing recognition of environmental degradation as a constraint on sustainable development. As the world's second-largest economy and a major energy consumer, China could no longer be a passive observer in global environmental politics.

The Reframing of Climate Policy

In June 2007, China released its first comprehensive policy blueprint—"China's National Climate Change Programme"—marking a strategic turning point in domestic governance and international engagement [11]. This was the first such plan issued by a developing country, articulating national goals for emissions reduction, energy efficiency, and adaptation. The document declared that addressing climate change was "not only an environmental necessity but an opportunity for technological and economic transformation." From 2008 onward, the Chinese government began publishing annual reports titled China's Policies and Actions for Addressing Climate Change, showcasing domestic progress and signaling policy transparency. These reports served both as tools of accountability and as instruments of soft power, reinforcing China's image as a responsible participant.

Diplomatic Activism and Constructive Mediation

China's diplomatic posture also became more assertive. In the 2009 Copenhagen Conference, China faced immense international pressure as the world's largest emitter of greenhouse gases. However, rather than retreating, Chinese negotiators adopted a dual strategy: defending the developmental rights of the Global South while promoting a pragmatic, bottom-up approach to global mitigation. Premier Wen Jiabao's personal involvement at the conference underscored the political weight Beijing attached to climate diplomacy. At Copenhagen, China voluntarily announced the goal of reducing carbon intensity per unit of GDP by 40–45% by 2020 compared with 2005 levels. Although not a legally binding commitment, this self-imposed target represented a significant shift from the purely defensive stance of the 1990s. In subsequent conferences—Bali (2007), Poznań (2008), Copenhagen (2009), and Cancun (2010)—China assumed the role of bridge-builder between developed and developing countries, advocating for inclusivity and fairness in the emerging post-Kyoto framework [12]. During the Durban Conference in 2012 (whose preparatory negotiations occurred in this decade), China proposed that all parties, including non-Kyoto developed states, should commit to comparable mitigation efforts. It also emphasized that developing countries' voluntary actions must be recognized under the UNFCCC, thereby ensuring both equity and participation. This pragmatic flexibility facilitated consensus and prevented the collapse of multilateral negotiations.

Domestic–International Nexus

Domestically, China linked climate governance to its broader modernization strategy. The Eleventh Five-Year Plan (2006–2010) incorporated energy efficiency and emissions intensity reduction as key indicators of performance. Between 2005 and 2010, China reduced energy intensity by 19.1% and invested heavily in renewable energy technologies. These efforts not only mitigated emissions but also catalyzed the emergence of a vibrant green economy, positioning China as a global leader in solar and wind manufacturing.

However, this proactive engagement also exposed new tensions. Internationally, developed countries argued that China's growing emissions made it a de facto great power that should bear greater responsibility. Domestically, policymakers faced the challenge of reconciling economic growth with environmental constraints. As one Chinese scholar noted, "To develop and to reduce emissions are no longer contradictory—they are two sides of the same modernization coin." [13] By the end of this decade, China had firmly repositioned itself as a proactive and open partner in the global climate regime—balancing national interests with global obligations, and shaping the agenda through both normative persuasion and material contribution.

5.3.3 The Leading Contributor and Normative Architect (2011–2021)

The decade following 2010 marked a decisive transition in China's climate diplomacy—from participation to leadership. This phase coincided with China's consolidation as a global economic powerhouse, the launch of the Belt and Road Initiative (BRI), and an increasing emphasis on ecological civilization as a guiding ideology. In this period, China did not merely implement global rules—it sought to shape them.

Institutional Leadership and the Paris Breakthrough

China played a pivotal role in the negotiations leading to the Paris Agreement (2015), the first universal and legally binding climate accord applicable to all countries. Through intensive diplomacy, especially with the United States, China helped bridge the gap between developed and developing countries. The 2014 China–US

Joint Announcement on Climate Change—in which both countries pledged quantitative targets for emissions control—paved the way for the Paris breakthrough [14]. At Paris, China supported a flexible architecture of Nationally Determined Contributions (NDCs), emphasizing that global success required self-differentiated yet collectively ambitious commitments. The resulting bottom-up model represented a new equilibrium between fairness and feasibility—a direct reflection of China's diplomatic philosophy of "pragmatic multilateralism."

Domestic Transformation and the "Dual-Carbon" Goals

The domestic translation of this leadership role was embodied in the 2030–2060 dual-carbon goals, first announced by President Xi Jinping at the 75th UN General Assembly in September 2020. China committed to peaking carbon emissions before 2030 and achieving carbon neutrality by 2060 [15]. This was not only a national policy but also a global pledge—aligning China's trajectory with the objectives of the Paris Agreement. To achieve these targets, China launched a series of structural reforms: establishing a national carbon trading market, accelerating renewable energy deployment, and integrating low-carbon objectives into the 14th Five-Year Plan. By 2020, carbon intensity had declined by 48.4% compared to 2005 levels, exceeding the voluntary commitment made in Copenhagen [16]. Moreover, China's leadership extended beyond its borders. The government initiated extensive South–South cooperation programs on climate change, providing financial aid, capacity-building, and technology transfer to developing countries. Over 200 training programs were organized, benefiting more than 5,000 officials from Asia, Africa, and Latin America [17]. Through these efforts, China repositioned itself as both a beneficiary and provider of global public goods.

Ideational Innovation: Ecological Civilization and Shared Destiny

The maturation of China's leadership also involved ideological innovation. The concept of Ecological Civilization (生态文明)—enshrined in the Communist Party Constitution in 2012—provided an indigenous framework linking domestic sustainability to global stewardship. Complementing this was the international discourse of a Community of Shared Future for Mankind, emphasizing inclusivity, cooperation, and mutual benefit [18]. Together, these frameworks constituted a distinctive form of normative leadership—one that contrasts with Western liberal environmentalism by prioritizing developmental justice and collective action. In this sense, China's leadership was less about domination and more about agenda-setting—offering what has come to be known as the China Solution (Zhongguo Fang'an): a model of governance that seeks harmony between growth and sustainability, sovereignty and cooperation.

Synthesis: From Participation to Leadership

Across these three decades, China's transformation in global climate governance has followed a discernible trajectory. The 1990s were marked by cautious engagement rooted in sovereignty and equity; the 2000s witnessed open participation grounded in pragmatic cooperation; and the 2010s culminated in proactive leadership characterized by institutional innovation and normative entrepreneurship. This evolution reflects both internal and external dynamics: the maturation of China's domestic governance capacity, the recalibration of global expectations, and the systemic redistribution of power in world politics. More than a story of compliance, it is a narrative of co-evolution—in which China and the global climate regime have mutually shaped each other's identities, priorities, and structures. As the next section will discuss, this transformation has produced substantial domestic achievements, new international partnerships, but also profound challenges that will determine the future trajectory of global climate governance.

5.4 Achievements, Challenges, and Strategic Measures

5.4.1 Domestic Achievements: Integrating Development and Decarbonization

The most striking feature of China's participation in global climate governance is the degree to which international commitments have been translated into concrete domestic transformation. Far from being a rhetorical exercise, China's climate policies have generated tangible achievements in emissions control, energy restructuring, and the construction of a national carbon governance architecture. These accomplishments represent not only the material basis of China's international credibility but also a paradigmatic case of developmental decarbonization—the simultaneous pursuit of economic modernization and environmental responsibility [19].

The Decoupling of Growth and Emissions

Since the 12th Five-Year Plan (2011–2015), China has consistently adopted carbon intensity—measured as CO_2 emissions per unit of GDP—as a binding indicator within national development plans. Between 2005 and 2020, carbon intensity fell by 48.4%, surpassing the target pledged at the Copenhagen Conference (40–45%) and avoiding an estimated 5.8 billion tons of cumulative CO_2 emissions [16]. This decline occurred alongside a fourfold increase in GDP, demonstrating that economic growth and emissions can, under effective policy coordination, be decoupled. The national policy framework that enabled this achievement combined regulatory and market-based instruments. Key mechanisms included the Energy Conservation Law (2008, revised 2018), sectoral efficiency standards, and pilot carbon trading schemes in Beijing, Shanghai, Guangdong, and other provinces. In July 2021, these pilots culminated in the launch of the National Emissions Trading System (ETS), which covers more than 4 billion tons of CO_2—making it the world's largest carbon market by volume [20]. China's success in lowering emissions intensity while maintaining robust growth has transformed it from an object of global scrutiny into an example of "green developmentalism." The underlying logic of this transformation is pragmatic: climate action is not a constraint on modernization but a new source of competitiveness, innovation, and legitimacy.

Restructuring the Energy Mix

Energy production and consumption lie at the heart of China's climate challenge. For decades, coal dominated the national energy mix, accounting for over 70% of primary energy use in the early 2000s. By 2020, this share had fallen to 56.8%, while the share of non-fossil energy rose to 15.9%, an increase of 8.5 percentage points since 2005 [21]. This restructuring reflects a systematic shift toward a multi-source, low-carbon energy architecture. Between 2010 and 2020, China's installed capacity of non-fossil generation reached 980 GW, representing 44.7% of total capacity. Hydropower, wind, solar, nuclear, and biomass energy experienced exponential growth—solar capacity increased by over 3,000 times compared with 2005, and wind capacity by 200 times [22]. Technological innovation and state-led industrial policy were the primary drivers of this transformation. The government's "Made in China 2025" strategy and the Renewable Energy Law (2006) fostered economies of scale in solar panel and wind turbine production, making China the world's largest producer and exporter of renewable technology. Meanwhile, investments in nuclear energy and smart-grid infrastructure have supported grid stability and integration of renewables. Such advances illustrate a distinctive path of state-guided energy transition, in which centralized planning, market mechanisms, and technological innovation operate synergistically. As a result, China has become both the world's largest consumer of renewable energy and a global supplier of clean energy technologies.

Green Mobility and Urban Transformation

The transportation sector, a major source of emissions, has also undergone profound restructuring. Between 2010 and 2020, China developed the world's most extensive network of high-speed railways, public transit systems, and electric vehicles. The number of new energy vehicles (NEVs) reached 5 million by 2020, accounting for more than half of the global total [23]. Government incentives—such as subsidies, license plate privileges, and charging infrastructure—created a self-reinforcing cycle of technological innovation and consumer adoption. Concurrently, the expansion of urban rail systems in 43 cities and the implementation of the "National Transit Metropolis" initiative enhanced public transportation capacity, reducing per capita emissions and congestion. These developments signify not only technological progress but also a redefinition of urban modernity: sustainable mobility has become a core component of China's vision of an "ecological civilization."

5.4.2 International Achievements: From Participant to Provider of Public Goods

China's climate achievements are not confined to its borders. Over the past decade, it has increasingly assumed the role of a provider of global environmental public goods through institutional participation, diplomatic leadership, and transnational cooperation.

Institutional Engagement and Normative Innovation

China's consistent commitment to multilateralism contrasts with the cyclical withdrawal and re-engagement of some Western powers. Unlike the United States, which exited and rejoined the Paris Agreement, China has demonstrated policy continuity and treaty adherence, thereby reinforcing institutional stability. As a "constructive stakeholder," it has actively participated in all Conferences of the Parties (COPs), contributed to negotiation outcomes, and supported the operationalization of the Paris framework [24]. Moreover, China has enriched global climate governance with new normative concepts. The idea of a "Community of Shared Future for Mankind" emphasizes mutual benefit and collective responsibility, challenging the zero-sum logic of traditional power politics. Similarly, the Ecological Civilization discourse bridges the divide between development and environmentalism by promoting harmony between humanity and nature [25]. These ideological innovations constitute a form of soft power diplomacy: they project an alternative vision of global order that aligns environmental protection with development justice. While some Western analysts interpret these frameworks as instruments of geopolitical influence, within China they are viewed as a moral response to the inequities of globalization.

South–South Cooperation and Green Belt and Road

Another dimension of China's international contribution is its promotion of South–South cooperation in climate governance. Since 2011, China has funded more than 200 capacity-building projects in developing countries, focusing on renewable energy, waste management, and disaster resilience [26]. The establishment of the China–UN Peace and Development Fund and the South–South Cooperation Fund on Climate Change has institutionalized these initiatives. Under the Belt and Road Initiative (BRI), China has increasingly prioritized green investment. In 2020, 57% of China's overseas energy investment flowed into renewable sectors—solar, wind, and hydro—compared with 38% in 2019 [27]. This marked a structural shift from fossil-fuel infrastructure toward low-carbon projects, aligning BRI financing with global sustainability norms. In 2021, President Xi Jinping announced that China would cease building new coal-fired power plants abroad, a move widely hailed as a "game changer" for global climate mitigation. Through these actions, China has repositioned itself as both a beneficiary and a benefactor of global environmental governance—helping to bridge the North–South divide and promoting a more equitable climate order.

5.4.3 Challenges and Structural Constraints

Despite these remarkable achievements, China's climate governance faces significant domestic, regional, and international challenges. These challenges stem from structural tensions within the global system as well as contradictions inherent in China's development model.

International Fragmentation and Geopolitical Tensions

The global climate regime remains fragmented by competing interests and declining multilateral cohesion. The resurgence of great-power competition, particularly between China and the United States, has re-politicized environmental cooperation. During the Trump administration, Washington withdrew from the Paris Agreement and framed climate governance as a burden rather than a shared responsibility. Although the Biden administration rejoined the accord, strategic rivalry has continued to undermine mutual trust [28]. European countries, while rhetorically committed to climate action, have faced crises—refugee influxes, Brexit, the COVID-19 pandemic, and the Russia–Ukraine war—that have diverted political capital from environmental leadership. The erosion of Western commitment has forced China to balance dual roles: sustaining global momentum while navigating geopolitical suspicion.

The Development–Decarbonization Dilemma

Domestically, the tension between economic development and emission reduction persists. Despite progress, coal remains the backbone of China's energy system, and regional disparities in economic structure complicate uniform implementation of low-carbon policies. Industrial provinces reliant on heavy manufacturing face social and economic costs from energy transition, raising concerns about employment and regional inequality [29]. Moreover, achieving the 2060 carbon neutrality goal demands technological breakthroughs in carbon capture, hydrogen energy, and energy storage—areas where China still depends partly on imported technologies. Without sufficient innovation and investment, the pathway to deep decarbonization could stall.

The Capacity Gap Among Developing Partners

As the leader of South–South cooperation, China also faces the expectation–capacity gap in assisting developing partners. Many recipient countries lack regulatory frameworks, financial instruments, or governance capacity to implement green projects effectively. Meanwhile, Western skepticism toward China's overseas investments sometimes constrains collaboration. Ensuring that the "Green Belt and Road" remains genuinely sustainable thus requires institutional safeguards and transparent governance.

Global Collective Action Problems

Finally, the fundamental dilemma of global climate governance remains unresolved. Collective action failures—free-riding, the tragedy of the commons, and the prisoner's dilemma—continue to undermine cooperation. Without enforceable global sanctions or incentive mechanisms, states may prioritize short-term national gains over long-term planetary stability. For China, navigating this anarchic environment necessitates both strategic flexibility and institutional creativity.

5.4.4 Strategic Measures and Forward Orientation

Facing these multidimensional challenges, China has adopted a combination of domestic reforms, regional cooperation, and global institution-building to sustain its leadership in climate governance.

The establishment of the Central Leading Group for Carbon Peaking and Carbon Neutrality in 2021 marked the institutional elevation of climate policy to the highest level of state governance. The 1 + N Policy Framework—with one master plan and multiple sectoral strategies—ensures coherent implementation across industries and provinces [30]. Within East Asia, China promotes subregional cooperation through mechanisms such as the ASEAN–China Partnership on Eco-Environmental Cooperation and the Trilateral Environmental Ministers' Meeting with Japan and South Korea. These frameworks help align regional standards and accelerate technology diffusion. Internationally, China advocates for a fair, rational, and win–win global climate order. By emphasizing equity, CBDR, and green recovery, China seeks to re-legitimize multilateralism and resist the re-politicization of environmental issues. Xi Jinping's Three Initiatives at the 2020 Climate Ambition Summit—"cooperation for mutual benefit," "a system of respective capabilities," and "green recovery"—encapsulate this approach [31]. China has become a pioneer in green finance, establishing one of the world's largest green bond markets. In partnership with the EU and international organizations, it has developed the Green Investment Principles for the Belt and Road, integrating environmental standards into overseas financing.

Through these multidimensional measures, China aims to consolidate its image as a responsible leader—one that not only honors commitments but also provides policy innovations to address systemic deficiencies in global governance.

China's achievements in climate governance—both domestic and international—demonstrate the country's capacity for strategic adaptation and normative innovation. Yet these successes coexist with profound structural tensions. Balancing national interests with global responsibilities, reconciling developmental imperatives with ecological constraints, and navigating a fragmented international environment all require continuous recalibration. The paradox of China's climate diplomacy lies in its dual identity: it is simultaneously a developing country defending its growth rights and a major power shaping global rules. The ability to harmonize these identities will determine whether China can sustain its leadership role in the next phase of global climate governance.

5.5 Conclusion: Toward a Fair and Cooperative Climate Future

Over the past three decades, China's journey in global climate governance has mirrored the transformation of the international order itself—from the hierarchies of a Western-centered system toward the pluralized, contested, and interdependent landscape of the twenty-first century. From a cautious participant defending developmental sovereignty in the 1990s, to a proactive partner mediating between North and South in the 2000s, and finally to a leading contributor shaping institutions and norms in the 2010s, China's trajectory reflects a continuous process of learning, adaptation, and innovation. Its experience reveals a fundamental truth: climate governance is not merely a technical matter of emission reduction, but a political project of justice, legitimacy, and shared responsibility. By embedding environmental ambition within its modernization strategy and linking national rejuvenation to global sustainability, China has demonstrated that ecological transformation can coexist with economic dynamism. Through the concepts of Ecological Civilization and the Community of Shared Future for Mankind, China has articulated a vision of global governance that transcends the old dichotomy of development versus environment, proposing instead a model of inclusive, cooperative, and fair transformation. Yet the road ahead remains arduous. Geopolitical fragmentation, technological bottlenecks, and structural inequalities continue to challenge the effectiveness of collective action. In navigating these complexities, China's role will remain pivotal—not as a hegemon imposing norms, but as a catalyst enabling consensus, innovation, and equity. As the world stands at the intersection of ecological crisis and institutional renewal, China's climate diplomacy offers a vital proposition: that the legitimacy of power in the twenty-first century will no longer rest on dominance, but on the capacity to build a sustainable future for all.

References

1. Yanhua L (2013) Promoting the construction of a harmonious and win-win global climate change governance system. Int Outlook 4:1
2. Wei S (2017) The overall national interest of a major power: a theoretical analysis. Mod Int RelatS 3:37–63
3. Waltz K (2003) Theory of international politics (trans: Xin Q). Shanghai People's Press, p 92
4. Yan X, Yan L (2008) Analysis of international relations. Peking University Press, pp 90–91
5. Xun P (2006) The definition and transformation of international role: a dynamic analytical framework. Int Polit Stud 1:134
6. Xiaoyong C (2023) China's role evolution in the process of global climate governance. Mod Int RelatS 6:124–145
7. Li XE (1991) Global environmental issues and China's principle position. China Popul Resour Environ (2):31–32
8. Luo JB (2007) Recollections on the negotiation and signing of the climate change framework convention. China Meteorol News, Jan 20, p 4
9. Haibin Z (2007) China and the international climate change negotiations. Int Polit Stud 1:21–36
10. Guiyang Z (2008) International climate governance in the post-Kyoto era and China's strategic choice. World Econ Polit 8:9
11. State Council of China (2007) National programme for addressing climate change. June 4
12. Hu J (2009) Joining hands to address the challenge of climate change. People's Daily, Sept 23, p 1
13. Li, Zhifei, Liang D, Haibin Z (2021) Thirty years of China's participation in international climate governance. China Popul Resour Environ 9:202–210
14. The White House (2014) US–China joint announcement on climate change. November 12

15. Xi J (2020) Statement at the general debate of the 75th session of the UN General Assembly. September 22
16. China's Policies and Actions for Addressing Climate Change (2022) Environ Prot (21):45–56
17. Ministry of Ecology and Environment of China (2021) China's South–South cooperation on climate change. Beijing
18. Yue S, Yu, Xiao (2019) China's promotion of global climate governance transformation from the perspective of a community of shared future for mankind. Northeast Asia Forum 6:112–123
19. Hui X (2023) The philosophical foundations and practical pathways of China's green development in the new era. Theor Inq 6:96–102
20. Ministry of Ecology and Environment of China (2021) China's national carbon emissions trading market report. Beijing
21. State Council of China, White Paper on Climate Change: China's Policies and Actions, October 2021
22. Bai SP (2023) China's green energy transition and its global impact. Mine Saf Inf (2)
23. Ministry of Transport (2019) China green transport development framework. People's Transport Press, Beijing
24. Song YM, Yu HY (2018) The Sino–US co-leadership structure in global climate governance: origins, stalemate, and reconstruction. Int Relat Stud (2):137–158
25. Dai JW (2023) Building a community of life between humanity and nature under global climate governance. Master's Thesis, Lanzhou University
26. Ministry of Ecology and Environment of China (2021) South–South cooperation on climate change annual report. Beijing
27. BP (2022) Statistical review of world energy 2022, 71st edn. BP Global, London
28. IEA (2023) World energy outlook 2023. International Energy Agency, Paris
29. Zhuang GY, Zhou ZG, Wang SB (2023) China's carbon neutral policy under the global green competition. Latin Am Stud (1)
30. State Council of China (2021) 1+N policy system for carbon peaking and carbon neutrality. Beijing
31. Xi J (2020) Three proposals for a new era of global climate cooperation. Address at the climate ambition summit, December 12

Chapter 6
Japan: Ambition, Innovation and the Search for Balance

As a major developed economy in East Asia, Japan has played a complex and shifting role throughout the thirty-year history of global climate governance. This chapter systematically examines the evolution of Japan's climate policies, their underlying motivations, and their outcomes from the early 1990s to the present. The analysis reveals that Japan's participation in climate governance has been driven by a combination of its strategic pursuit of political power, energy security realities, domestic experience with natural disasters, and changes in the international landscape. Its trajectory exhibits distinct phases, transitioning from an "Active Explorer" to an "Ambitious Frontrunner," and then to a "Cautious Participant." Although Japan once cultivated a leading environmental image through its hosting of the Kyoto Protocol, its environmental ODA, and its technological energy efficiency, its leadership and credibility have faced severe challenges following the Fukushima nuclear accident, its increasingly conservative stance in international negotiations, and structural contradictions within its domestic emission reduction efforts. Finally, this chapter explores the opportunities and uncertainties for Japan in reshaping its climate leadership through energy structure transformation and technological innovation under its carbon neutrality goal.

6.1 Drivers and Foundations: Internal and External Motivations for Japan's Participation in Climate Governance

Japan's engagement with global climate governance is not an isolated diplomatic act but an inevitable choice shaped by the interplay of its national strategy, resource endowment, historical experience, and international circumstances. This system of motivations forms the cornerstone for understanding the logic behind Japan's climate actions.

X. Zou, *Thirty Years of Global Climate Governance*, Energy, Climate Change & Sustainability in the Global South, https://doi.org/10.1007/978-981-95-8457-4_6

6.1.1 Regulation and Pull of the International System

The formation and development of the global climate governance regime exerted external "regulatory power" and "normative pull" on Japan. In the late 1980s, the climate change issue rapidly gained prominence within international scientific and political agendas. The signing of the United Nations Framework Convention on Climate Change (UNFCCC) in 1992 marked the initial establishment of a global climate governance framework, enshrining the principle of "common but differentiated responsibilities." The formation of this global consensus provided the initial international context and institutional framework for Japan's participation in climate governance [1].

The adoption of the Kyoto Protocol in 1997 was a milestone in the history of international climate governance. It not only set internationally legally binding emission reduction targets but also innovatively introduced three flexible implementation mechanisms: Joint Implementation (JI), the Clean Development Mechanism (CDM), and International Emissions Trading (IET). Crucially, its birth in Kyoto, Japan, thrust the nation onto the center stage of global climate governance. As the host country, Japan demonstrated its diplomatic influence in facilitating the agreement, and the Kyoto Protocol became a key platform for Japan to enhance its international status and export its environmental理念 and technologies [2]. Subsequent Conferences of the Parties (COPs) continuously refined and deepened the governance mechanisms, creating an external field of pressure and opportunity that Japan had to continually respond to and engage with. Japan's climate governance exhibits a significant "internationally-driven, domestically-responsive" characteristic, where the evolution of international institutions acts as a major catalyst for its domestic policy legislation and strategic adjustments.

6.1.2 Rigid Domestic Constraints and Transition Drivers

The fundamental internal drivers for Japan's participation in climate governance stem from its severe energy security situation and frequent natural disasters.

The Imperative of Energy Security: Japan is an archipelagic nation extremely poor in energy resources, heavily reliant on imports for fossil fuels. According to data from Japan's Ministry of Economy, Trade and Industry (METI) 2023 Energy White Paper, its energy self-sufficiency rate has long hovered at low levels. In the 1960s, relying on domestic coal, the self-sufficiency rate reached 58.1%. As the economy shifted primarily to oil during high-speed growth, the rate plummeted. Although the introduction of nuclear power brought some improvement, the 2011 Fukushima nuclear accident reduced nuclear power generation to nearly zero, causing the energy self-sufficiency rate to hit a historical low of 6.3% in FY2014. Despite subsequent recovery due to renewable energy development and nuclear restarts, the rate stood at only 13.3% as of FY2021 [3]. This high external dependency makes the Japanese economy highly vulnerable to fluctuations in the international energy market, as evidenced by the two oil crises. Therefore, promoting energy transition, developing renewable energy, and advancing energy-saving technologies are not merely about emission reduction; they are a matter of survival concerning the nation's economic lifeline and security.

The Acute Pain of Natural Disasters: Japan is located on the Pacific Ring of Fire and has a temperate maritime monsoon climate, making it prone to frequent natural disasters such as earthquakes, volcanoes, typhoons, and heavy rainfall. According to statistics from the Global Disaster Data Platform, between 2013 and 2021, Japan experienced over 70 disasters including tropical storms, floods, and earthquakes [4]. The increased frequency of extreme weather events and sea-level rise due to climate change pose a direct and severe threat to Japan. This geographical vulnerability objectively heightens Japanese society's sensitivity and sense of crisis regarding climate change, fostering greater "resonance" when the international community calls for climate action.

6.1.3 Legacy of Historical Experience and Strategic Tool Conversion

Japan's climate governance did not start from scratch. The painful lessons and successful experiences from its domestic pollution control provided crucial "experiential capital" and a source of "legitimacy" for its participation in global environmental governance. The post-WWII economic-priority development model once turned Japan into a "pollution powerhouse," with shocking incidents like Minamata disease and Itai-itai disease. In response, Japan underwent profound reflection, successfully managing industrial pollution through strict legislation, technological innovation, and public participation, and gradually established a sustainable development path aimed at building a recycling-oriented society [5]. This domestic transformation from "polluter" to "manager" endowed Japan with a degree of persuasiveness and practical experience when promoting its environmental理念, technologies, and cooperation models to the international community.

At a deeper level, climate governance was strategically positioned by Japan as a springboard for achieving its "political power" aspirations. After the Cold War, with its economic strength consolidated, the pursuit of political power status became Japan's core national strategy. However, as a WWII defeated nation, its actions in "high politics" areas (such as military and security) were highly constrained by its domestic constitution and the international community. Consequently, "low politics" issues like environment and climate became an important突破口 for expanding its international influence and shaping an image of a "responsible major power." [6] By leading regional environmental cooperation, providing environmental Official Development Assistance (ODA), and actively participating in international climate regimes, Japan sought to accumulate leadership capital in the environmental field, which could then be translated into broader political influence, gradually establishing its discursive power in Asia–Pacific and global affairs.

6.2 Process and Evolution: Japan's Triple Role Transformation in Climate Governance

Looking back at Japan's climate actions over the past three decades, its role has not been static but has undergone significant shifts—from an explorer, to a frontrunner, and then to a cautious participant—in response to changes in the domestic and international political and economic environment.

6.2.1 Exploration and Foundation-Laying: An Active Explorer in Global Climate Governance (Late 1980s—End of 20th Century)

This phase marked Japan's period of recognizing and initially engaging with global climate governance, characterized by both international responsiveness and domestic foundation-laying.

At the international level, Japan began incorporating environmental issues into its diplomatic purview. In 1988, Prime Minister Noboru Takeshita expressed concern about global environmental degradation at the G7 Toronto Summit, an early signal of Japanese leadership's attention to climate issues. In 1992, Japan signed the UNFCCC, becoming a significant signatory. Crucially, at COP3 held in Kyoto in 1997, Japan, as the host, used its diplomatic mediation to facilitate the adoption of the Kyoto Protocol, instantly establishing its special status in global climate governance [7]. Concurrently, Japan's environmental diplomacy began taking shape at the regional level, initiating the construction of an Asia–Pacific environmental cooperation network through measures such as hosting the Asia–Pacific Environmental Conference, implementing environmental ODA in Southeast Asia, and establishing the Northeast Asian Conference on Environmental Cooperation (NECC).

Domestically, Japan began building the institutional framework for climate governance. It established the Ministerial Conference on the Global Environment as a policy coordination platform. Legislatively, the 1993 Basic Environment Law macroscopically established national environmental protection policies, and the 1998 Law Concerning the Promotion of Measures to Cope with Global Warming provided specific legal basis for addressing climate change. During this stage, Japan used international commitments to compel domestic institutional development, but its domestic governance system, academic research, and public participation remained nascent, with emission reduction pathways still unclear.

6.2.2 *Ambition and Leadership: An Ambitious Frontrunner in Global Climate Governance (Early 21st Century—2010)*

Entering the twenty-first century, against the backdrop of the Kyoto Protocol's entry into force, Japan entered an "ambition phase" in climate governance, striving to become a leader through multi-pronged strategies.

First, deepening energy conservation and emission reduction. Japan pushed energy-efficient technology to the extreme, promoting "smart energy conservation" and "technical energy conservation." Through series of regulations like the Act on the Rational Use of Energy (the Energy Conservation Act), combined with information technology to precisely identify high-consumption sectors, and continuously improving vehicle fuel efficiency and promoting efficient appliances, these measures made Japan one of the most energy-efficient countries globally, providing a solid foundation for meeting its targets under the Kyoto Protocol's first commitment period (2008–2012).

Second, proposing ambitious governance strategies. Despite frequent prime ministerial changes during this period, successive leaders consistently prioritized climate issues, proposing landmark initiatives. These included Junichiro Koizumi's "Koizumi Initiative" promoting the Kyoto Protocol's entry into force; Shinzo Abe's "Cool Earth 50" plan during his first administration; Yasuo Fukuda's "Cool Earth Promotion Initiative" and "Fukuda Vision"; and Yukio Hatoyama's ambitious mid-term target announced at the 2009 UN Climate Summit to reduce emissions by 25% by 2020 compared to 1990 levels [8]. While focusing on different aspects, these strategies collectively signaled Japan's strong desire to lead the global emission reduction wave.

Third, expanding the breadth and depth of environmental diplomacy. Japan's environmental ODA became a key tool for expanding its influence (See Fig. 3: Japan's Environmental ODA Budget and Priority Areas). Globally, it consolidated its "environmental powerhouse" image through financial and technological exports. Regionally, institutionalized cooperation strengthened further, with platforms like the "Japan-ASEAN Dialogue on Environmental Cooperation" allowing Japan to gradually assume a leading role in Asia–Pacific environmental affairs. By constructing an extensive "network of environmental governance partnerships," Japan effectively enhanced its international soft power.

6.2.3 *Shock and Recalibration: A Cautious Participant in Global Climate Governance (2011—Present)*

The Great East Japan Earthquake and the Fukushima nuclear accident in 2011 marked a major turning point in Japan's climate governance process. The confluence of a series of domestic and international factors led to a shift towards a cautious, even conservative, stance.

The Fukushima accident led to the prolonged shutdown of the vast majority of nuclear power plants in Japan, causing a drastic regression in the energy structure. To compensate for the power shortfall, Japan was forced to significantly increase imports and power generation from liquefied natural gas (LNG) and coal (See Fig. 5: Changes in Japan's Overall Energy Supply Structure [1990–2020]), resulting in a rebound in greenhouse gas emissions and a setback for emission reduction progress. Concurrently, the "Abenomics" policy package introduced after Shinzo Abe returned to power in 2012 prioritized economic growth, implementing deregulation in environmental areas, which weakened the policy drive for emission reduction. Faced with rising energy costs and future policy uncertainty, corporate confidence in emission reduction was generally low [9].

Against this backdrop, Japan's position in international negotiations significantly backtracked. At the 2010 Cancun climate conference, Japan publicly rejected a second commitment period under the Kyoto Protocol; at the 2012 Doha conference and the 2019 Madrid conference, it was twice awarded the "Fossil of the Day" prize by the Climate Action Network (CAN) for its pro-coal stance. This contrasted sharply with its first-phase leader image.

However, it is noteworthy that Japan did not entirely abandon climate governance domestically. It announced a "2050 Carbon Neutrality" goal and, through documents like the Green Growth Strategy and the Basic Policy for Achieving GX (Green Transformation), sought to promote innovation in areas like hydrogen and renewable energy. This contradictory stance of being "hot internally, cold externally" precisely reflects Japan's difficult trade-offs between energy security, economic competitiveness, and climate responsibility. Therefore, defining Japan in this phase as a "Cautious Participant" rather than simply an "Obstructor" is more accurate, as its actions are filled with realist calculations.

6.3 The Core and the Key: Japan's Energy Policy Transformation Under the Carbon Neutrality Goal

The essence of climate governance is energy governance. Whether Japan can achieve its climate goals fundamentally depends on the deep decarbonization of its energy system. Under the 2050 carbon neutrality pledge, Japan's energy policy is undergoing a profound transformation.

6.3.1 *Evolutionary Context: From "3E" to "3E + S"*

The evolution of Japan's energy policy is a microcosm of its response to domestic and international challenges. Post-WWII, it transitioned from coal to "oil-centered, coal-supplemented." Following the two oil crises, Japan initiated its first strategic shift focused on energy conservation and diversification (developing nuclear power, natural gas), enacting landmark laws like the Energy Conservation Act (1979). In the 1990s, against the rising climate agenda, policy goals shifted from solely "Energy Security" to "3E Coordination," balancing Energy Security, Economic Efficiency, and Environmental protection [10].

The Fukushima nuclear accident was another turning point. The cornerstone of energy policy added "Safety" to the "3E" foundation, forming the "3E + S" principle. This meant nuclear restarts had to pass the world's strictest safety standards, and energy policy formulation needed to prioritize facility operational safety. Guided by this, Japan formulated its 4th to 6th Strategic Energy Plans, entering a new phase of diversification and decarbonization premised on safety.

6.3.2 *Strategic Blueprint: The 6th Strategic Energy Plan and the 2030/2050 Vision*

The 6th Strategic Energy Plan, released in 2021, serves as Japan's roadmap towards carbon neutrality. It explicitly positioned renewables as the "main power source" for the first time and set the 2030 energy mix target: renewables 36%-38%, nuclear power 20%-22%, and fossil fuels (equipped with CCUS etc.) about 41% [11]. This target reflects Japan's balancing act between ideal and reality.

a) Prioritizing Renewable Energy Development: Japan strives to overcome bottlenecks like limited land area and geographical constraints by transitioning from a Feed-in Tariff (FIT) to a Feed-in Premium (FIP) scheme, optimizing the grid, and reducing costs to vigorously promote solar, wind (especially floating offshore wind). However, high generation and system costs (See Fig. 9: Changes in Renewable Energy Cost Burden in Japan) and limited suitable land and sea areas remain barriers to large-scale expansion.
b) The Controversial Return of Nuclear Power: Facing the dual challenges of low energy self-sufficiency and emission reduction pressure, the Japanese government views restarting nuclear plants meeting new safety standards as an indispensable transitional option for ensuring baseload power and achieving the 2030 target. The GX Decarbonization Power Supply Act passed in 2023 even removed the principle 40-year operational limit for nuclear plants, allowing extensions beyond 60 years with regulatory approval [12]. While controversial domestically, this move reflects Japan's choice under practical constraints.
c) Strategic Bet on Hydrogen and Ammonia: Japan regards hydrogen as the trump card for achieving carbon neutrality by 2050. Since formulating the Basic

Hydrogen Strategy in 2017, through state-led R&D, infrastructure subsidies (e.g., hydrogen stations), and international supply chain building, it aims to create a world-leading "hydrogen society."[13]. Key focuses include supporting hydrogen/ammonia co-firing power generation and developing hydrogen reduction steelmaking technology. However, currently high hydrogen costs and insufficient "green hydrogen" scale are major obstacles to commercial rollout.

d) Energy Conservation and Grid Modernization: Japan continually revises the Energy Conservation Act, expanding the definition of "energy" to all areas, and deepening conservation measures in industrial, commercial, and residential sectors. Simultaneously, it promotes the development and application of smart grids and storage battery technology to enhance grid stability and the integration of intermittent renewables.

6.3.3 Motivations and Challenges: Green Growth and Structural Constraints

The drivers for Japan's energy transition are multidimensional. Beyond the fundamental needs of addressing climate change and ensuring energy security, creating new economic growth momentum is a significant consideration. The administrations of both Yoshihide Suga and Fumio Kishida have positioned green transformation as central to their "New Capitalism," estimating that the "Green Growth Strategy" could generate approximately 190 trillion yen in annual economic benefits and numerous jobs by 2050 [14]. This indicates Japan's intent to turn the climate challenge into an opportunity for industrial upgrading and economic development.

However, Japan's energy transition faces a difficult path ahead. First, structural constraints on the energy mix are prominent. Path dependency on fossil fuel imports is hard to shake off in the short term (See Fig. 8: Changes in Japan's Fossil Fuel Dependency), with thermal power remaining the ballast for power stability. Second, techno-economic challenges are severe. Renewables, hydrogen, nuclear power, and other areas all face cost, technological, or social acceptance issues. Finally, international competition is intensifying. In the global race for new energy technology and rule-making dominance, Japan faces fierce competition from China, the United States, and Europe, challenging its traditional technological advantages.

6.4 The Path to Reshaping Japan's Climate Leadership

Japan's three-decade journey in climate governance is a chronicle of seeking balance between ambition and dilemma. Its experiences and lessons hold significant mirroring significance for global climate governance.

6.4.1 Re-Examining the Historical Experience

Japan's practice demonstrates that integrating climate governance into the core national development strategy can generate powerful momentum. Through its early "Environmentally Sustainable Nation" strategy, it successfully transformed the negative legacy of pollution control into a positive asset for environmental diplomacy, and leveraged its leading advantages in energy-saving and environmental technology to shape a favorable international image. Simultaneously, the Japanese case highlights the importance of a sound legal system and long-term mechanisms. From the Basic Environment Law and the Global Warming Countermeasures Promotion Law to the various Strategic Energy Plans, although policies fluctuated with changes in government, the basic legal framework and continuously advancing energy conservation and emission reduction efforts maintained considerable continuity.

However, Japan's lessons are equally profound. First, the vulnerability of the energy structure is the greatest systemic risk. The shock to energy policy triggered by the Fukushima nuclear accident directly led to the interruption of its emission reduction progress and the backtracking of its international stance. This illustrates that without a robust, clean, and independent energy supply system, a nation's climate ambition is like building a castle on sand. Second, maintaining international credibility requires consistent action. The shift from being a promoter of the Kyoto Protocol to a rejecter of its second commitment period severely damaged Japan's credibility and moral authority as an "environmental powerhouse." Third, domestic consensus is the ballast for policy. For highly contentious issues like nuclear power and carbon pricing, a lack of broad social consensus and transparent decision-making processes often makes policy implementation difficult.

6.4.2 Future Prospects: Challenges and Opportunities Coexist

Looking ahead, Japan's path in climate governance remains fraught with challenges. The international trust crisis triggered by the decision to discharge treated water from the Fukushima plant [15] has further damaged its environmental image. Domestically, energy-intensive industries show insufficient willingness to face deep emission reduction targets, and corporate confidence in emission reduction needs revitalization. Furthermore, in the context of the increasing "high politicization" of global climate governance, Japan's strategy of using climate issues to enhance its international status faces a complex situation characterized by intensified major power rivalry and the collective rise of emerging nations.

Nevertheless, opportunities still exist. The irreversible global trend towards carbon neutrality provides a new arena for Japan, which possesses technological reserves in areas like energy efficiency, hydrogen, and storage batteries. If it can successfully transform its technological advantages into industrial and standard-setting advantages, Japan is poised to secure a place in the future green economy competition. The key lies in whether Japan can demonstrate genuine political courage to undertake a thorough energy revolution, overcome entrenched interests, significantly increase the share of renewable energy, and build a more resilient and sustainable energy system.

6.5 Conclusion

From Rio to Glasgow, the changes in Japan's role on the global climate governance stage vividly the complex dance of a middle power pursuing political ambition, ensuring economic security, and fulfilling environmental responsibilities. It was once crowned with glory through the Kyoto Protocol, and later overshadowed by the Fukushima accident and conservative positions. Its journey tells us that climate leadership stems not only from ambitious commitments but, more crucially, depends on the resolve to overcome domestic structural contradictions, consistent international credibility, and the wisdom to find a way out through technological innovation and global cooperation. For Japan, the path to reshape its climate leadership is precisely a national transformation journey to break the energy curse and achieve green revival. The success or failure of this path concerns not only Japan itself but will also provide important insights for many other nations facing similar challenges.

References

1. United Nations (1992) United Nations framework convention on climate change
2. United Nations Framework Convention on Climate Change. (1998) Report of the conference of the parties on its third session, held at Kyoto from 1 to 11 December 1997. FCCC/CP/1997/7/Add.1
3. METI (2023) Energy white paper 2023. Ministry of Economy, Trade and Industry, Japan
4. Global Disaster Data Platform (2023) Disaster event query. https://www.gddat.cn/. Accessed 18 Nov 2023
5. Mizuguchi S (2012) Japan's post-Cold War environmental diplomacy strategy. Northeast Asia Forum 21(3):98–104
6. Wang Y, Ge J, Wu Z (2017) An analysis of Japan's global governance strategy. Contemp Asia-Pac 5:26–50, 157
7. United Nations Framework Convention on Climate Change (1998) Kyoto Protocol to the United Nations Framework Convention on Climate Change
8. Bi Z (2019) A study on Japan's participation in global climate governance. China Foreign Affairs University, PhD diss.
9. Du J, Yutong Q (2022) The dilemma and realization path of Japan's move towards 'carbon neutrality.' Mod Jpn Econ 41(3):66–80
10. Piao Y, Hu, Xiyue (2023) Motivations and implementation paths of Japan's energy transition from the perspective of carbon neutrality. Mod Jpn Econ 42(4):14–27
11. Ministry of Economy, Trade and Industry (2021) The 6th strategic energy plan. Ministry of Economy, Trade and Industry, Japan

12. Wei R (2023) Japan's energy transition and nuclear power choice. Energy 4:52–55
13. Li H (2021) The experience of Japan's 'hydrogen society' construction and its enlightenment to China. Jpn Res 4:33–42
14. Cabinet Office (2021) Green growth strategy through achieving carbon neutrality in 2050. Cabinet Office, Japan
15. Jin Y (2023) The formation, development, and deep-seated causes of the Fukushima treated water issue. Jpn Stud 2:34–55, 153–154

Chapter 7
India's Climate Path: Balancing Development Rights with Environmental Accountability

Over the past three decades, India's trajectory in global climate governance has embodied the quintessential dilemmas of the Global South: the constant negotiation between the right to development and the responsibility to mitigate climate change. As an emerging power with over 1.4 billion people, an expanding industrial base, and growing energy demand, India has become an indispensable actor in shaping the future of the planet's climate regime. Its position has never been static. Rather, it has evolved dynamically in response to shifting geopolitical contexts, domestic imperatives, and scientific awareness. Understanding India's climate politics therefore offers not only an insight into one nation's policy evolution, but also a window into the broader transformation of developing countries' engagement with global environmental governance.

From the Rio Earth Summit in 1992 to the Glasgow Conference in 2021, India's role has evolved from that of a defensive advocate of the principle of "common but differentiated responsibilities" to a pragmatic participant seeking to reconcile its development imperatives with the imperatives of decarbonization. The evolution of this stance can be divided roughly into four distinct yet interconnected phases: (1) the period of principled resistance during the 1990s; (2) the gradual shift toward pragmatic flexibility in the 2000s; (3) the emergence of proactive engagement under the Paris Agreement framework; and (4) the recent articulation of long-term "net zero" ambitions.

X. Zou, *Thirty Years of Global Climate Governance*, Energy, Climate Change & Sustainability in the Global South, https://doi.org/10.1007/978-981-95-8457-4_7

7.1 From Principle to Pragmatism: The Evolution of India's Position

7.1.1 The Era of Principles and Defensive Diplomacy (1990s—early 2000s)

During the 1990s, when the international architecture for climate governance was first being constructed through the United Nations Framework Convention on Climate Change (UNFCCC, 1992) and later the Kyoto Protocol (1997), India's overriding priority was clear: to safeguard its development sovereignty. The country had only recently embarked upon economic liberalization, and millions remained mired in poverty. For Indian policymakers, any international commitment that constrained fossil fuel consumption or industrial growth was viewed as a potential obstacle to poverty eradication and modernization.

At the heart of India's diplomatic stance was an unyielding defense of fairness. The principle of *common but differentiated responsibilities and respective capabilities* (CBDR-RC) became its normative cornerstone. Indian negotiators, drawing upon early conceptual work by the Centre for Science and Environment (CSE), articulated the idea of "per capita entitlements"—the notion that each human being has an equal right to the global atmospheric commons, and that the historical overuse of this space by industrialized nations constituted a form of environmental colonialism [1]. In successive Conferences of the Parties (COPs)—Rio (1992), Berlin (1995), Kyoto (1997)—India played a leading role in embedding this principle in the text of the UNFCCC. While the Kyoto Protocol introduced mechanisms such as joint implementation, emissions trading, and the Clean Development Mechanism (CDM), India remained cautious, fearing that such flexibility might dilute the moral boundary between developed and developing countries and shift the burden of mitigation southwards.

Domestically, this defensive posture reflected the broader ethos of postcolonial developmentalism: economic growth was synonymous with national sovereignty. As official statements repeatedly emphasized, "economic development and poverty eradication are India's overriding priorities." Environmental commitments that threatened to impede these goals were rejected as unjust intrusions by the Global North. Consequently, during the 1990s India was frequently portrayed by Western commentators as a "hardliner" or even an "obstructionist" actor in climate negotiations—a perception rooted less in hostility than in a profound structural asymmetry of development stages [2, 3].

7.1.2 Emerging Awareness and Strategic Adjustment (Mid-2000s—Copenhagen)

Entering the new millennium, two intertwined forces began to erode India's purely defensive orientation: the growing domestic recognition of climate vulnerability, and intensifying international pressure for emerging economies to assume greater responsibility. The publication of the IPCC's Fourth Assessment Report in 2007 profoundly shaped India's policy discourse. It warned that South Asia would face more frequent extreme weather events, glacier retreat, reduced agricultural productivity, and severe water stress [4]. For India, whose monsoon-dependent agriculture and densely populated coastlines made it acutely exposed, climate change was no longer an abstract global concern but a tangible national security issue. The idea that adaptation and resilience should become integral to India's development planning began to take hold.

In response, the government launched the **National Action Plan on Climate Change (NAPCC)** in 2008—a comprehensive framework of eight "national missions" encompassing solar energy, energy efficiency, sustainable habitats, water resources, Himalayan ecosystems, reforestation, sustainable agriculture, and knowledge creation [5]. This marked a decisive turning point: climate change was now embedded within the national development agenda, not treated as an external constraint. At the same time, India began to recalibrate its international posture. In 2007, Prime Minister Manmohan Singh declared that India's per capita greenhouse gas emissions "will never exceed those of developed countries" [6]. This voluntary pledge, while maintaining the per capita logic, opened the door to future commitments. It reflected a strategic shift—an effort to preserve moral legitimacy while acknowledging growing global expectations.

The Copenhagen Conference of 2009 (COP15) served as a crucible for this evolving pragmatism. India joined with China, Brazil, and South Africa to form the **BASIC coalition**, a powerful bloc that mediated between the demands of industrialized nations and the concerns of the developing world. While India continued to reject legally binding quantitative targets, it voluntarily committed to reducing the *emissions intensity* of its GDP by 20–25% from 2005 levels by 2020 [7]. Furthermore, India proposed the mechanism of "International Consultation and Analysis" (ICA)—a compromise formula balancing the transparency demanded by developed countries with the sovereignty concerns of developing ones [8].

The transformation was subtle but profound. India was no longer the intransigent "nay-sayer" of the Kyoto years; it was emerging as a **constructive broker**, seeking solutions that preserved equity while accommodating the realities of a rapidly changing global order.

7.1.3 The Paris Moment and the Reconfiguration of India's Role

The election of Narendra Modi in 2014 ushered in a new phase of climate diplomacy marked by self-confidence and a desire for global leadership. Modi's government viewed environmental stewardship as compatible with national resurgence—a discourse that fused civilizational pride with technological modernity. In the lead-up to the 2015 Paris Climate Conference, India submitted an ambitious **Nationally Determined Contribution (NDC)** committing to a 33–35% reduction in emissions intensity from 2005 levels by 2030, and pledging that 40% of its electricity generation capacity would come from non-fossil fuel sources [9]. The swift ratification of the Paris Agreement signaled India's acceptance of a universal, "bottom-up" regime—one that preserved differentiation while granting flexibility for national circumstances.

Yet India's Paris moment was not merely about compliance; it was about leadership. Together with France, India co-founded the **International Solar Alliance (ISA)**, a coalition of over 120 sun-rich countries aimed at mobilizing investment and technology for solar energy deployment [10]. The ISA symbolized a shift from reactive participation to proactive agenda-setting. Alongside this, India launched other global initiatives such as the *Coalition for Disaster Resilient Infrastructure (CDRI)* and the *One Sun, One World, One Grid* vision, positioning itself as a thought leader on sustainable development in the Global South.

This reorientation represented more than diplomatic rebranding. It reflected India's recognition that climate governance could serve as an arena for soft power projection and geoeconomic opportunity. By championing solar energy and resilience infrastructure, India sought not only to meet domestic needs but also to shape the normative vocabulary of global climate politics—one grounded in equity, inclusiveness, and technological partnership rather than coercive conditionality.

7.1.4 Toward Net Zero: Glasgow and Beyond

By the time of the 2021 Glasgow Conference (COP26), the global conversation had shifted decisively toward long-term net-zero targets. Under mounting pressure to articulate a clear decarbonization trajectory, Prime Minister Modi announced that **India would achieve net zero emissions by 2070**—a historic commitment that nonetheless reflected India's developmental realities [11]. India simultaneously updated its near-term targets, pledging to reduce the emissions intensity of GDP by 45% by 2030 and to ensure that half of its electricity generation capacity would come from non-fossil sources [12]. While critics argued that the 2070 timeline was less ambitious than the 2050 or 2060 goals set by other major emitters, the announcement nonetheless represented a milestone in India's climate narrative. For the first time, the country explicitly linked its domestic transition strategy to the global goal of carbon neutrality.

Over thirty years, India's stance thus underwent a profound metamorphosis: from the rigid defense of historical responsibility to the embrace of voluntary mitigation; from a posture of resistance to one of participation; and from a reactive recipient of rules to an active contributor to their design. Beneath this evolution lies a consistent thread—the pursuit of *equitable responsibility*—yet its expression has become progress India and Global Climate Governance: Balancing Development Rights and Climate Responsibility.

7.2 India and Global Climate Governance: Balancing Development Rights and Climate Responsibility

If India's international posture over the past three decades reveals a gradual shift from defensive resistance to pragmatic engagement, its domestic transformation tells an equally compelling story. Climate governance within India has evolved from a peripheral concern to a central pillar of national policy. This transition has been driven not only by external diplomatic commitments but also by internal imperatives: energy security, economic modernization, and the rising social costs of climate vulnerability. The institutionalization of climate policy in India illustrates a complex balancing act between federal and state authorities, between growth and sustainability, and between the competing demands of mitigation and adaptation.

India's approach to climate governance has been characterized by a layered institutional framework designed to coordinate across ministries and levels of government. The creation of the Prime Minister's Council on Climate Change in 2007 was a landmark moment[13]. Chaired by the Prime Minister himself, the Council brought together ministers from key portfolios—environment, energy, finance, agriculture, and science—along with representatives from civil society and academia. This high-level mechanism ensured that climate considerations were no longer confined to the Ministry of Environment and Forests (later renamed the Ministry of Environment, Forest and Climate Change, MoEFCC) but were integrated into the national planning process. The Council's role extended beyond policy formulation to monitoring implementation and facilitating inter-ministerial coordination.

Parallel to this, specialized agencies were created to operationalize specific components of the climate agenda. The National Clean Development Mechanism Authority (2003) was among the first, facilitating India's active participation in the Kyoto Protocol's CDM market [14]. Over time, this evolved into a broader ecosystem of climate-related institutions—from the Bureau of Energy Efficiency (BEE) overseeing industrial energy standards to the Energy Efficiency Services Limited (EESL), a public-sector company driving large-scale efficiency projects.

The institutional design reflects a recognition that climate governance cannot be siloed within environmental policy. It requires integration with economic planning, industrial policy, and fiscal management. Yet, the challenge has always been implementation. India's federal structure means that much of the execution rests with state governments, whose capacity and political will vary widely. As such, the national architecture often functions as a framework for guidance and coordination, while actual progress depends heavily on subnational dynamics.

7.3 The Policy Framework: From Strategy to Action

The cornerstone of India's domestic climate policy is the National Action Plan on Climate Change (NAPCC) launched in June 2008 [5]. It articulated an ambitious vision: to achieve sustainable economic growth while addressing climate challenges through eight "national missions." Each mission was led by a different ministry and focused on a key sector or cross-cutting issue—solar energy, energy efficiency, sustainable habitats, water, Himalayan ecosystems, reforestation, sustainable agriculture, and strategic knowledge for climate change. Among these, two missions emerged as particularly influential: the National Solar Mission and the National Mission on Enhanced Energy Efficiency (NMEEE). Together, they exemplify India's dual strategy of pursuing low-carbon growth through both supply-side and demand-side reforms.

7.3.1 The National Solar Mission: From Symbol to System

Launched in 2010 under the NAPCC, the Jawaharlal Nehru National Solar Mission (JNNSM) set a bold target of deploying 20 GW of solar power by 2022—a goal later expanded to 100 GW, and subsequently integrated into the national renewable energy target of 500 GW by 2030[15]. The mission's design combined visionary ambition with pragmatic economics. Through competitive bidding, feed-in tariffs, and a transparent regulatory framework, the government succeeded in attracting substantial private investment and driving down the cost of solar energy. Between 2010 and 2022, the levelized cost of solar electricity in India fell by over 85%, making it one of the cheapest sources of power globally. The success of the solar mission was not merely technological; it was institutional and social. By promoting decentralized solar projects—rooftop systems, solar pumps for agriculture, and mini-grids in remote villages—the mission aligned clean energy with rural development. Solar parks in states like Rajasthan, Gujarat, and Karnataka became symbols of India's green modernity, demonstrating how renewable energy could contribute simultaneously to growth, employment, and environmental stewardship.

7.3.2 The Perform, Achieve and Trade Mechanism: Marketizing Efficiency

On the demand side, the Perform, Achieve and Trade (PAT) scheme launched in 2012 represented a major innovation in energy efficiency policy [16]. Modeled on a cap-and-trade framework, it set specific energy consumption reduction targets for energy-intensive industries—including steel, cement, power generation, and fertilizer—and allowed firms that overperformed to sell "energy saving certificates" to those that underperformed. The PAT scheme demonstrated how market-based instruments could reconcile environmental objectives with industrial competitiveness. By the end of its first cycle, it had achieved energy savings equivalent to around 31 million tonnes of CO_2. The second and third cycles expanded coverage to additional sectors, gradually embedding efficiency considerations into corporate management practices.

Meanwhile, EESL spearheaded large-scale efficiency programs, notably the Ujala initiative, which distributed over 360 million LED bulbs nationwide, reducing annual electricity consumption by more than 47 billion kWh. These initiatives showcased how centralized procurement and economies of scale could deliver both economic and environmental dividends.

7.3.3 The Hydrogen Frontier and Emerging Technologies

As the global energy transition accelerates, India has moved to position itself at the forefront of next-generation technologies. In 2021, it launched the National Green Hydrogen Mission, envisioning the country as a hub for the production and export of green hydrogen [17]. This initiative aligns with India's twin goals of reducing fossil fuel dependency and building a globally competitive green manufacturing base. Hydrogen, alongside nuclear power and energy storage, is viewed as a strategic component of India's low-carbon future. The government continues to promote nuclear energy—a stable, low-emission source capable of complementing the variability of renewables. Despite international restrictions in the past, India has developed indigenous reactor technologies and expanded its nuclear fleet under the "Make in India" industrial strategy.

7.3.4 Adaptation and Climate Resilience

While mitigation dominates international discourse, adaptation remains an existential imperative for India. Recognizing this, the government created the National Adaptation Fund for Climate Change (NAFCC) in 2015 to support state-level projects addressing agriculture, water, and ecosystem resilience [18]. Investments in climate-resilient infrastructure, drought-resistant crops, and early warning systems have become integral to development planning. States such as Odisha and Gujarat have pioneered local adaptation strategies, particularly in coastal management and disaster preparedness. The devastating cyclones and heatwaves of recent years have underscored the urgency of these efforts, demonstrating that the cost of inaction far exceeds that of proactive adaptation.

7.4 Energy Transition in Practice: Three Illustrative Case Studies

To understand the tangible progress and challenges of India's energy transformation, it is instructive to examine specific examples where policy ambition met real-world implementation. The following three cases—in biomass energy, wind power, and nuclear development—illustrate the diversity of India's approaches and the interplay of technological, economic, and social factors.

7.4.1 Biomass Energy and Rural Circular Economy: The Gudutur Power Plant

India's agricultural sector generates vast quantities of residues such as straw and bagasse, much of which has traditionally been burned in the open, contributing to severe air pollution. Recognizing this, the government has encouraged biomass utilization as a means of combining energy generation with waste management and rural employment. A notable example is the Gudur Power Plant in Tamil Nadu, developed by EID Parry, which uses locally sourced sugarcane bagasse and crop residues as feedstock. Equipped with circulating fluidized bed (CFB) boiler technology, the plant produces approximately 320,000 kWh annually with an installed capacity of 40 MW [19].

The project's impact is multidimensional:

- Energy Contribution: It provides electricity for roughly 3,000 households, improving local energy reliability.
- Environmental Benefit: By preventing stubble burning, it reduces seasonal smog and achieves measurable carbon reductions.
- Socioeconomic Value: It creates direct employment for dozens of workers and offers farmers additional income through residue sales, establishing a localized circular economy.

Government incentives, including preferential feed-in tariffs, have been crucial to scaling such projects. Nationally, biomass-based installed capacity more than doubled between 2015 and 2022, underscoring its potential as a decentralized renewable resource [20].

7.4.2 Scaling Wind Power: The Muppandal Wind Farm

With its long coastline and high wind potential, India ranks among the world's largest wind energy producers. The Muppandal Wind Farm in Tamil Nadu epitomizes this success. Developed through a consortium of private and public investors, the project now hosts turbines with a cumulative capacity exceeding 300 MW [21]. Muppandal's significance lies not only in its size but in the industrial ecosystem it has nurtured. The project catalyzed domestic manufacturing of wind turbines, blades, and components, reducing reliance on imports and fostering local employment. The economies of scale achieved through competitive auctions have dramatically lowered costs, making wind power one of India's most cost-effective energy sources. By integrating wind farms into regional grids and upgrading transmission networks, India has demonstrated how large-scale renewable projects can deliver both environmental gains and industrial spillovers. Today, wind accounts for nearly one-third of India's renewable capacity, supporting its broader goal of energy diversification.

7.4.3 Nuclear Energy and Technological Sovereignty: The Kalpakkam Plant

India's pursuit of nuclear power reflects a long-term commitment to energy independence and technological self-reliance. The Kalpakkam Nuclear Power Station in Tamil Nadu, operated by the Nuclear Power Corporation of India Limited (NPCIL), represents a cornerstone of this strategy [22]. What distinguishes Kalpakkam is its indigenous design—the Pressurized Heavy Water Reactor (PHWR)—developed and built entirely by Indian scientists and engineers. This project symbolizes India's ability to master the full nuclear fuel cycle, from design and construction to fuel fabrication and waste management. Strategically, nuclear energy provides stable base-load power to complement intermittent renewables. It also carries geopolitical significance: India's perseverance in developing civilian nuclear capabilities despite decades of technology denial has reinforced its status as a technologically sovereign state. While safety concerns and high capital costs remain, the Kalpakkam case highlights India's determination to secure low-carbon energy through multiple pathways—not by abandoning traditional technologies, but by indigenizing and improving them.sively more sophisticated and pragmatic.

The impressive strides India has made in climate policy and renewable energy expansion coexist with daunting structural challenges. These are not merely technical or financial obstacles, but deeply embedded social, economic, and institutional dilemmas. The very scale and diversity that make India a potential leader in the global energy transition also complicate its decarbonization pathway. Understanding these constraints is essential to appreciating both the achievements and the limits of India's climate governance.

7.5 Structural Challenges

7.5.1 The Coal Paradox and Energy Transition

At the core of India's energy dilemma lies the paradox of coal. Despite its massive investment in renewables, coal continues to supply over half of India's total energy consumption [23]. The reasons are structural and historical: coal is abundant, cheap, and deeply interwoven into the political economy of industrialization. Millions of livelihoods depend on coal mining and associated sectors, particularly in states such as Jharkhand, Chhattisgarh, and Odisha. Rapidly phasing out coal is therefore not merely a technological issue but a social one. The transition to clean energy entails managing "just transition" concerns—re-skilling workers, diversifying regional economies, and compensating for fiscal losses in coal-dependent states. While India's renewable energy expansion has been remarkable, it has largely added new capacity rather than replacing old thermal plants. The coexistence of "incremental greening" and "locked-in carbon intensity" remains one of the central contradictions of India's energy landscape. In policy terms, the challenge is compounded by infrastructure bottlenecks. Renewable energy generation often outpaces grid modernization. Transmission losses, inadequate storage capacity, and regulatory fragmentation across states hinder full utilization of renewable output. The Indian government has recognized this and launched major grid reforms, including the Green Energy Corridor initiative and investments in battery storage, yet the scale of the problem is vast.

7.5.2 Financing and Technology Gaps

Achieving India's NDCs and long-term net-zero targets will require investments estimated at several trillion U.S. dollars by mid-century [24]. While private capital has flowed into solar and wind projects, financing for adaptation, grid resilience, and industrial decarbonization remains insufficient. Domestic banks often view clean energy projects as risky due to policy uncertainty and long payback periods. India has consistently emphasized that international climate finance commitments—particularly the promised USD 100 billion per year from developed countries—have not been fully met. The gap between rhetoric and reality has fostered skepticism in New Delhi about the credibility of global commitments. As a result, India has turned increasingly toward blended finance mechanisms, green bonds, and multilateral development bank partnerships to mobilize resources.

Technology transfer poses a similar challenge. Key decarbonization technologies—such as carbon capture and storage (CCS), advanced battery systems, and green hydrogen electrolysers—remain dominated by developed countries. Intellectual property barriers and limited local manufacturing capacity constrain India's ability to scale up at the required pace. While India has made progress in indigenous innovation, the absence of a robust global framework for technology sharing continues to impede equitable progress.

7.5.3 Governance Complexity and Policy Coordination

India's federal system, while a source of democratic resilience, often complicates the coherence of climate policy. Energy and environmental issues straddle multiple ministries and agencies. Overlapping mandates and jurisdictional ambiguities can lead to fragmented decision-making. For instance, renewable energy promotion falls under the Ministry of New and Renewable Energy (MNRE), but land acquisition, grid management, and industrial policy involve other ministries and state authorities. The result is a mosaic of policy instruments that do not always align seamlessly. Some states—such as Gujarat, Tamil Nadu, and Maharashtra—have become pioneers in renewable energy deployment, while others lag behind due to political, fiscal, or infrastructural constraints. The unevenness of implementation underscores the need for stronger intergovernmental coordination and capacity-building at the subnational level.

Moreover, procedural democracy and environmental litigation, though vital to transparency, can sometimes delay major projects. Public consultation requirements and judicial interventions have strengthened environmental accountability but occasionally produced policy paralysis. Balancing democratic participation with administrative efficiency remains an enduring tension.

7.5.4 Climate Impacts and the Cost of Vulnerability

Even as India invests in mitigation, the tangible impacts of climate change continue to escalate. The country ranks among the most climate-vulnerable nations in the world. Recurrent floods, droughts, heatwaves, and cyclones have inflicted severe human and economic losses. The 2023 monsoon season, for instance, saw catastrophic flooding in Himachal Pradesh and Uttarakhand, while heatwaves in northern India led to record-breaking temperatures exceeding 47 °C. These disasters impose a double burden. On the one hand, they divert fiscal resources from long-term decarbonization toward short-term relief and reconstruction. On the other hand, they disproportionately affect the poorest populations, reinforcing the developmental inequalities that underpin India's cautious stance in international negotiations. The contest between "adaptation needs" and "mitigation imperatives" reflects a deeper issue: climate justice within and beyond national borders.

7.6 Prospects and Pathways

Despite these challenges, India's future in global climate governance is one of expanding agency rather than retreat. The structural constraints it faces do not negate its potential for leadership; they define the contours within which leadership must be exercised. Several trends point toward a more assertive and sophisticated role for India in the coming decades.

7.6.1 From Participant to Co-Architect of Global Rules

India's evolution from a reactive participant to a co-architect of global climate governance is already visible. Its initiatives such as the International Solar Alliance (ISA) and the Coalition for Disaster Resilient Infrastructure (CDRI) have redefined how developing countries can shape multilateral agendas. Unlike traditional diplomacy focused on defensive negotiations, these initiatives reflect a proactive strategy of norm entrepreneurship—setting the agenda, framing discourses, and building coalitions around shared technological and developmental priorities. In the post-Paris landscape, where the governance model is increasingly bottom-up and polycentric, India's influence will depend on its ability to connect domestic achievements with international leadership. Demonstrating credible progress in renewable energy, energy efficiency, and adaptation will enhance its moral and political leverage. At the same time, India will continue to advocate for differentiated responsibilities and equitable transitions, positioning itself as the voice of the Global South.

7.6.2 Deepening the Domestic Green Transformation

At home, the next phase of India's climate strategy will involve a deeper integration of climate objectives with industrial and social policy. The government's Production-Linked Incentive (PLI) schemes for solar manufacturing, electric vehicles, and batteries indicate a shift from policy targets to industrial ecosystems. This approach seeks to transform climate action into an engine of economic growth and job creation. The ongoing expansion of renewable capacity—targeting 500 GW by 2030—will likely be accompanied by accelerated investment in storage technologies and smart grids. Hydrogen and biofuels may play complementary roles in decarbonizing hard-to-abate sectors such as steel, cement, and transport. The policy discourse has increasingly adopted the language of "energy independence" (Atmanirbhar Bharat), linking decarbonization with national self-reliance. Equally important will be the integration of social justice into the transition process. The government's plans for "just transition" roadmaps in coal regions reflect a growing recognition that the energy shift must be inclusive. Ensuring that green growth delivers equitable benefits—across classes, castes, and regions—will be critical for maintaining political legitimacy.

7.6.3 Pragmatic International Cooperation and South–South Solidarity

India's external strategy will likely continue to balance principled advocacy with pragmatic engagement. While maintaining pressure on developed countries to fulfill their finance and technology transfer commitments, India will also deepen cooperation with emerging economies. The BASIC group (Brazil, South Africa, India, China) remains an important platform for coordinating negotiating positions within the UNFCCC, but India has also cultivated diverse partnerships through the G20, BRICS, and Indo-Pacific frameworks. South–South cooperation—in renewable technology sharing, climate finance, and adaptation planning—will be a growing dimension of India's diplomacy. Its experience in delivering low-cost solar and wind solutions positions it to assist other developing countries in achieving their own energy transitions. Moreover, collaboration with China, despite geopolitical tensions, holds significant potential in areas such as grid modernization, hydrogen technology, and carbon market development. By leveraging its developmental experience and technological capacity, India can redefine what leadership means in the climate arena—not as dominance, but as collaborative stewardship.

7.6.4 Policy Recommendations for a Sustainable Trajectory

Looking ahead, several priorities emerge for strengthening India's climate governance:

- Institutional Integration: Establish a unified national climate authority to streamline policy coordination across ministries and states. Such an entity could consolidate climate-related funding mechanisms, monitoring frameworks, and reporting systems.
- Financial Innovation: Expand the use of green bonds, carbon pricing, and sovereign guarantee schemes to mobilize private investment. Domestic financial institutions should develop climate risk assessment tools to guide lending.

- Technological Breakthroughs: Increase public investment in R&D for storage, hydrogen, and carbon utilization. Partnerships between national laboratories and private industry could accelerate commercialization.
- Subnational Empowerment: Build technical and fiscal capacity at the state level to ensure effective implementation of national missions. Incentivizing inter-state competition in renewable deployment could yield positive spillovers.
- Adaptation Prioritization: Elevate adaptation planning to the same strategic level as mitigation. Developing integrated urban resilience strategies and climate-proof infrastructure should become a national imperative.
- Global Engagement: Actively shape emerging mechanisms on carbon markets, loss and damage, and climate finance transparency to safeguard equity and access for developing nations.

7.7 Conclusion

India's thirty-year journey in global climate governance is, in essence, a chronicle of transformation—from defensive sovereignty to strategic engagement, from principle-based opposition to pragmatic innovation. The country's evolving stance has mirrored its developmental trajectory: as its economy grew and its global aspirations expanded, so too did its capacity and willingness to assume greater responsibility. Yet, this evolution has never entailed abandoning the moral foundation of its climate diplomacy—the insistence on fairness, historical accountability, and differentiated responsibility. What makes India's experience distinctive is the dual logic that underpins it. On one hand, the persistence of poverty, energy insecurity, and social inequality anchors its demand for policy space. On the other hand, the growing domestic awareness of climate risks and the economic promise of green technologies propel it toward proactive leadership. The tension between these two logics—protection and participation—has defined India's negotiation strategy and domestic policymaking alike.

Domestically, India's achievements are tangible. The expansion of renewable energy, the institutionalization of climate governance, and the innovation of policy instruments such as the PAT scheme and solar missions testify to a significant shift in developmental priorities. Yet, the road to deep decarbonization remains long. Structural dependence on coal, fiscal constraints, and governance fragmentation continue to impede progress. Internationally, India's role is poised to grow in significance. Its ability to bridge the interests of developed and developing countries, to translate moral argument into institutional design, and to anchor its diplomacy in credible domestic action will determine its influence in the decades ahead. In many ways, India has become both a mirror and a microcosm of the global challenge: how to reconcile growth with sustainability, sovereignty with solidarity, and equity with ambition. Ultimately, the success of global climate governance will depend on whether the international system can accommodate the developmental realities of nations like India. A fair, effective, and inclusive regime—one that recognizes differentiated capacities while fostering cooperative innovation—is indispensable for global progress. India's choices, therefore, will not merely shape its own trajectory; they will profoundly affect the collective capacity of humanity to navigate the Anthropocene.

In the century-long journey toward carbon neutrality, India stands as both a test case and a torchbearer—embodying the hopes, contradictions, and potential of the developing world's engagement with the climate challenge. Its path, marked by negotiation and innovation alike, will remain a decisive variable in the unfolding history of global climate governance.

References

1. Centre for Science and Environment (1991) Global warming in an unequal world: a case of environmental colonialism. Centre for Science and Environment, New Delhi
2. United Nations (1992) United Nations framework convention on climate change. United Nations, New York
3. Jacobsen S (1998) India's position on climate change from Rio to Kyoto: a policy analysis
4. IPCC (2007) Climate change 2007: impacts, adaptation and vulnerability. Contribution of working group II to the fourth assessment report of the intergovernmental panel on climate change. Cambridge University Press, Cambridge
5. Government of India (2008) National action plan on climate change. Prime Minister's Council on Climate Change, New Delhi
6. Ministry of External Affairs (2007) PM's intervention on climate change at the Heiligendamm meeting. Government of India, New Delhi
7. Government of India (2010) India's submission to the UNFCCC regarding the Copenhagen accord. Ministry of Environment and Forests, New Delhi
8. Michaelowa K, Michaelowa A (2011) India in the international climate negotiations: from traditional nay-sayer to dynamic broker. CIS Working Paper (70)
9. Government of India (2015) India's intended nationally determined contribution. Ministry of Environment, Forest and Climate Change, New Delhi
10. International Solar Alliance (n.d.) About ISA. https://isolaralliance.org/. Accessed 13 June 2024
11. Prime Minister's Office (2021) National statement by Prime Minister Shri Narendra Modi at COP26. Government of India, New Delhi
12. Government of India (2022) India's updated nationally determined contribution. Ministry of Environment, Forest and Climate Change, New Delhi
13. Government of India (2007) Constitution of the Prime Minister's council on climate change. Cabinet Secretariat, New Delhi
14. Government of India (2003) Notification for the establishment of the National CDM Authority. Ministry of Environment and Forests, New Delhi
15. Ministry of New and Renewable Energy, Government of India. (n.d.) Physical progress (achievements). https://mnre.gov.in/. Accessed 2 Mar 2026
16. Bureau of Energy Efficiency, Government of India (n.d.) Perform, achieve and trade (PAT) Scheme. https://beeindia.gov.in/. Accessed 2 Mar 2026
17. Ministry of New and Renewable Energy (2021) National green hydrogen mission. Government of India, New Delhi
18. Government of India (2015) National adaptation fund for climate change. Ministry of Environment, Forest and Climate Change, New Delhi
19. Case study compiled from project documentation and industry reports. EID Parry is a publicly listed company in India

20. Ministry of New and Renewable Energy (2023) Annual report 2022–23. Government of India, New Delhi
21. : Tamil Nadu Generation and Distribution Corporation (TANGEDCO) (n.d.) Project information
22. Shukla U, Bajpai R (2019) Nuclear power plants in India: Achieving clean and green energy. Int J Nucl Energy Sci Technol 13(1):87–97
23. BP (2023) Statistical review of world energy 2023. BP, London
24. International Energy Agency (2021) India energy outlook 2021. IEA, Paris

Chapter 8
Brazil: Bridging Environmental Stewardship and Economic Growth

As the custodian of the planet's largest tropical rainforest and a pivotal voice from the Global South, Brazil's trajectory through three decades of global climate governance presents a profound and instructive narrative. This chapter examines Brazil's journey from a sovereignty-conscious observer to a period of constructive leadership, through a phase of stark regression, and into its current potential for renewed global engagement. It argues that the nation's climate diplomacy has been fundamentally shaped by the persistent tension between its domestic development imperatives and its escalating global environmental responsibilities. By tracing this evolving role, the chapter illuminates the complex interplay of geopolitics, economic ambition, and ecological urgency that defines not only Brazil's past but also critically informs its future path and, by extension, the prospects for global climate stability.

8.1 Introduction

Situated in the southeastern part of South America, Brazil stands as the continent's largest nation, spanning a vast territory of 8.51 million square kilometers. It shares borders with nearly every other South American country except Chile and Ecuador and boasts an extensive Atlantic coastline. Endowed with exceptional natural wealth, Brazil is home to the Amazon River—the world's largest river by discharge volume and basin area—and harbors unparalleled biodiversity, abundant arable land, a highly competitive agricultural sector, the world's third-largest freshwater reserves, and immense hydropower potential [1]. As the nation containing the largest portion of the Amazon rainforest and wielding significant regional influence, Brazil's impact on the Amazon is profound and direct. Consequently, the international community often regards Brazil as the "primary responsible actor" for the Amazon's protection [1]. Its domestic policies, particularly those governing forest exploitation and conservation,

X. Zou, *Thirty Years of Global Climate Governance*, Energy, Climate Change & Sustainability in the Global South, https://doi.org/10.1007/978-981-95-8457-4_8

not only determine the nation's ecological fate but also significantly influence the trajectory of global climate change.

As a developing country, Brazil has been a prominent actor on the stage of global climate governance. Frequently positioning itself as a mediator between the developing and developed worlds in international climate negotiations, Brazil has strategically employed climate diplomacy as a vital instrument to advance its "major power aspirations." Therefore, the evolution of Brazil's role in global climate governance over the past three decades is, in essence, a history of its continuous struggle to balance domestic development imperatives with its perceived global environmental responsibilities.

8.2 Brazil's Evolving Role in Global Climate Cooperation

Brazil's stance towards global climate cooperation has undergone a discernible trajectory, shifting from cautious observation to proactive leadership, followed by a period of wavering and retrenchment.

Brazil has consistently expressed a desire, and often a dependence, for financial and technological support from developed nations, while typically resisting any external interference in its domestic affairs [2]. Prior to the 1990s, Brazil endured a prolonged period of authoritarian rule, during which the military regime viewed environmental concerns as a conspiracy fabricated by industrialized nations to maintain their advantage [3]. Adhering to a doctrine that "development must not be sacrificed for environmental quality," the government pursued aggressive natural resource exploitation [4]. A telling example occurred in 1948 when UNESCO attempted to establish an international institute for scientific research and education in the Amazon region. This initiative ultimately failed, garnering support only from Ecuador and France. Brazil opposed the plan during negotiations, fearing that granting other nations equal rights in Amazonian affairs would lead to the "internationalization" of the region and compromise its national sovereignty.

A significant shift occurred in 1992 when Brazil hosted the United Nations Conference on Environment and Development (UNCED) in Rio de Janeiro. This landmark event, the largest and highest-level international conference on environment and development since the 1972 Stockholm Conference, marked a turning point. As the host nation, Brazil became the first signatory to the United Nations Framework Convention on Climate Change (UNFCCC), adopting a constructive and proactive stance in the ensuing negotiations. It made pivotal contributions, such as proposing the concept of a Clean Development Fund and emphasizing the "historical responsibility" of developed nations in emissions reduction. By launching voluntary emission reduction plans and mediating between developing and developed countries, Brazil leveraged its unique position to shape the foundational principles of the nascent global climate governance regime, seeking to secure greater influence on the climate issue.

The election of Luiz Inácio Lula da Silva of the Workers' Party in 2003 heralded a golden era for Brazilian climate diplomacy. Confronted with the escalating impacts of climate change, which severely affected the Amazon and Northeast regions, threatening local ecosystems, food security, and the livelihoods of Indigenous and traditional communities, the Lula administration integrated sustainable development into national policy. Domestically, it adopted a more proactive approach towards the Amazon, making concrete commitments to reduce deforestation and emissions while bolstering scientific research. In 2005, Brazil submitted a proposal to the UNFCCC Secretariat for the creation of an international fund designed to incentivize and reward efforts to reduce deforestation rates. This initiative played a crucial role in advancing international negotiations on what would later become the REDD + (Reducing Emissions from Deforestation and Forest Degradation) mechanism.

Internationally, Brazil's climate diplomacy became inextricably linked to its "major power ambitions." This strategy was threefold: first, to assume a leadership role in global climate governance and cultivate an image of a responsible global power, thereby elevating its international status; second, to steadfastly defend national sovereignty and development rights, asserting control over the Amazon while representing and uniting developing countries to secure their interests; and third, to promote its clean energy expertise, particularly in biofuels, transforming technological prowess into tangible economic competitiveness. As former Brazilian Environment Minister Rubens Ricupero aptly stated, "Brazil has everything it takes to be an environmental power, and this is the only field where we can achieve our aspiration to be a great power."

Brazil championed the inclusion of poverty alleviation as a central focus of the UN Millennium Development Goals, rallying international support to address inequality and poverty in Latin America. It forged coalitions with other rainforest-rich nations and developed countries to integrate REDD + into the UNFCCC negotiations and established the Amazon Fund, attracting international financial assistance for sustainable rainforest conservation. During the Warsaw climate talks in 2013, Brazil insisted on acknowledging historical responsibilities for climate change, using concrete actions and technical data to help lay the groundwork for the post-2020 global climate agreement [5]. It actively promoted its clean energy technologies across various multilateral forums on climate and energy, seeking not only to lead global emission reduction efforts but also to foster a worldwide market for bioenergy. Brazil also amplified its voice through the BRICS bloc. As consensus and willingness for climate cooperation grew among members, BRICS gradually established a climate cooperation mechanism underpinned by green finance, guided by energy transition goals, and implemented through new energy and environmental collaborations [6]. The New Development Bank (NDB), established by the BRICS nations, emerged as a key institution, providing innovative green financing and investment opportunities focused on infrastructure, thereby supporting member states' transitions to green economies and emission reductions.

Although the subsequent Dilma Rousseff administration, grappling with economic downturn, accorded less priority to environmental issues and witnessed some regression—such as encouraging carbon-intensive industries—it largely maintained the fundamental climate negotiation stance inherited from the Lula government.

However, the presidency of Jair Bolsonaro (2019–2022) marked a stark reversal, drawing widespread international criticism for its regressive climate and environmental policies. Prioritizing economic growth above all else, the Bolsonaro government dramatically departed from the proactive posture of the Workers' Party era. A member of the far-right Social Liberal Party and often dubbed the "Trump of Brazil," Bolsonaro campaigned on a platform threatening Brazil's immediate withdrawal from the Paris Agreement, arguing that compliance imposed undue economic costs and infringed upon national sovereignty, particularly over the Amazon. Upon taking office, his government promptly announced Brazil's withdrawal as host of the COP25 climate summit in 2019, citing budget constraints and its revised stance on climate change. At the rescheduled COP25 in Madrid, Brazil, represented by then-Environment Minister Ricardo Salles, contributed to a negotiating impasse by refusing to compromise with the European Union on Article 6 of the Paris Agreement. Concurrently, the administration slashed environmental budgets, weakened environmental agencies, and relaxed regulations governing forest exploitation and protected areas, including Indigenous territories, national and state parks, and conservation units.

Guided by these policies, deforestation in the Amazon surged by 34% in 2019, followed by a further 7% increase in 2020. Although Brazil's National Institute for Space Research (INPE) projected a slight decrease of 1–2% in deforestation for 2021, this still equated to over 7,000 square kilometers of forest loss in the first nine months alone. Defying the global trend of a 7% drop in greenhouse gas (GHG) emissions in 2020 due to the COVID-19 pandemic, Brazil's emissions soared by 9.5%, reaching their highest level since 2006, with deforestation hitting a 2008 high. Consequently, Brazil became one of the few major economies whose pollution levels did not decline during the pandemic. In a notable shift ahead of COP26 in October 2021, the Bolsonaro government launched the "National Program for Green Growth," outlining climate goals backed by approximately $400 billion in public and private investments aimed at cutting carbon emissions, conserving forests, and promoting sustainable resource use. An interministerial committee was established to oversee the plan, and at COP26, Brazil pledged to enhance its 2030 GHG emission reduction target from 43 to 50%. President Bolsonaro, in a pre-recorded message, declared, "Brazil is a part of the solution to climate change," signaling a dramatic, if largely superficial, change in rhetoric that yielded minimal tangible results [7].

The re-election of Luiz Inácio Lula da Silva in 2023 has set the stage for a decisive return to proactive climate policies. Having campaigned on strong environmental pledges, Lula faces high international expectations to rebuild the environmental governance framework systematically dismantled by his predecessor. Early signs, including a reported decrease in Amazon deforestation compared to the previous year, suggest that Brazil, under Lula's leadership, is poised to re-engage constructively in the global climate governance process.

8.3 Dilemmas and Challenges in Brazil's Climate Governance

Despite its potential and past initiatives, Brazil's path in global climate governance is fraught with persistent and complex challenges.

Foremost among these is the fundamental tension between economic development and environmental conservation. Although Brazil's per capita GDP ranks relatively high among developing nations and its industrial structure shares some features with developed economies, a significant gap in overall economic output and living standards remains. The Brazilian economy has oscillated between periods of rapid growth and profound stagnation over the past two decades. With the OECD projecting a slowdown in GDP growth from 2.5% in 2022 to 0.8% in 2023, the nation's economic outlook appears precarious. Challenges such as high inflation, unemployment, fiscal deficits, public debt, and low productivity growth plague its economy. This economic vulnerability undermines the financial foundation for sustained environmental protection, often forcing the government to prioritize immediate economic recovery over long-term ecological sustainability. Unlike many industrialized nations where emissions primarily stem from fossil fuel combustion, over 50% of Brazil's greenhouse gas emissions originate from Land Use, Land-Use Change, and Forestry (LULUCF), predominantly due to Amazon deforestation. Thus, the nation faces a cruel dilemma: the very resource base that could fuel its development—the rainforest—is being destroyed in the process, forcing this emerging "BRICS" economy to navigate an arduous developmental transition where "development versus conservation" remains a central, unresolved question.

The profound influence of domestic politics constitutes another major challenge. The commitment to climate action has varied dramatically with changing administrations, revealing a fragile political consensus. The Bolsonaro presidency (2019–2022) exemplified this volatility, actively dismantling environmental protections and oversight bodies, resulting in a 75.5% increase in the average annual deforestation rate compared to the previous decade. In stark contrast, President Lula's return in 2023 signaled a renewed commitment, with his government swiftly reinstating the legal framework for the Action Plan for the Prevention and Control of Deforestation in the Legal Amazon (PPCDAm) and reinforcing laws against environmental crimes. This stark policy swing based on electoral outcomes creates uncertainty and impedes long-term strategy. Furthermore, deeply entrenched corruption within public life continues to pose a significant risk, potentially diverting resources and undermining the effective implementation of environmental policies.

Societal and security factors also present formidable obstacles. A strand of nationalist sentiment, particularly potent during the Bolsonaro era, fosters a perception of the Amazon as a purely domestic affair, viewing any foreign interest or involvement as an infringement on national sovereignty. This mindset can hinder international cooperation and fuel resistance to global oversight. Additionally, Brazil's alarmingly high crime rates directly impact environmental governance. The peril faced by environmental activists and defenders is severe; reports indicate that over 440 environmental activists were killed in Brazil between 2002 and 2013, creating a climate of intimidation that stifles civil society oversight and grassroots conservation efforts.

Externally, Brazil's international credibility on climate issues has suffered significant damage. The Bolsonaro government's policies earned Brazil a reputation as a "climate villain" or "rogue state" in environmental forums, leading to a loss of trust among international partners and potential donors. Rebuilding this diplomatic capital and restoring its reputation as a reliable actor in global climate negotiations remains a critical challenge for the current administration.

8.4 Prospects for Brazil's Future Role in Global Climate Governance

Notwithstanding the challenges, there are compelling reasons for a cautiously optimistic outlook regarding Brazil's capacity to re-emerge as a constructive leader in global climate governance. The nation possesses the necessary resources, technological expertise, and, under the current government, the political will to strengthen its climate action. Its participation is widely recognized as indispensable for achieving global emission reduction targets.

The Lula administration has already taken steps to reactivate its climate diplomacy. A prominent example is the China-Brazil Joint Statement on Addressing Climate Change issued in April 2023, which reaffirmed the principle of Common but Differentiated Responsibilities (CBDR), pledged deepened bilateral cooperation, and expressed joint support for eliminating global illegal logging and deforestation [8]. Such partnerships are crucial for bolstering Brazil's international standing and access to resources.

Brazil's advanced bioenergy sector, particularly its pioneering role in sugarcane-based ethanol, represents a significant strategic asset. The country has accumulated substantial technical experience and expresses a willingness to share its biomass technology and resources internationally [9], positioning itself as a key player in the global transition to sustainable biofuels. Furthermore, Brazil is exploring frontier clean energy opportunities, with government studies underway for low-carbon hydrogen production projects estimated to have a potential output of 1.8 billion tons, attracting significant investment interest.

Domestically, Brazil has established a policy framework to incentivize its renewable energy sector. The government has implemented a series of fiscal incentives, such as the Renewable Energy Development Policy, which exempts infrastructure projects for renewable energy from import and federal taxes, and the Infrastructure Tax Exemption Policy, providing tax breaks for companies investing in renewable energy infrastructure [10]. These measures are designed to stimulate green investment and accelerate the energy transition.

In conclusion, Brazil stands at a critical juncture. The return of a government committed to environmental issues has opened a window of opportunity for the country to reclaim a central role in the global fight against climate change. While deep-seated economic pressures and political volatility pose enduring risks, the combination of renewed leadership, proven technological capabilities in clean energy, active international engagement, and supportive domestic policies suggests that Brazil's future in global climate governance can be viewed with cautious optimism. Its journey will be a critical test of whether a major developing economy can successfully harmonize its developmental aspirations with its pivotal global environmental responsibilities.

References

1. Zhou L (2020) Brazil's participation in Amazon regional governance: process and role analysis. Unpublished master's thesis, China Foreign Affairs University
2. Chen L (2022) Factors influencing the effectiveness of climate aid: a comparative study of Brazil's rainforest protection pilot project and the Amazon fund. J Lat Am Stud 44(1):85–102
3. Zhou L (2023) Shifts in Brazil's Amazon policy against the backdrop of US-EU climate leadership competition. J Lat Am Stud 45(1):134–152
4. Zuo P, Jiang P (2017) An analysis of the motivations and cooperation mechanisms of BRICS countries in global climate governance. Int Rev 4:57–71
5. Zhao B (2019) Brazil and South Africa in global climate politics: historical process, drivers of change, and identity selection. Foreign Theor Trends 4:74–86
6. Shen C (2019) BRICS cooperation under the adjustment of Brazil's foreign policy. World Aff 21:18–19
7. Yang J (2021) Is Brazil's new climate pledge and "green growth plan" a political stunt? The Paper. https://www.thepaper.cn/newsDetail_forward_15603376. Accessed 21 Oct 2023
8. Beijing Review (2023) China-Brazil joint statement on addressing climate change. Beijing Review 66(19):Insert 14–Insert 16
9. Wen X (2022) Status of Brazil's energy industry and opportunities for energy transport. China Shipp 5:68–73
10. Zhao J, Feng J, Yang Y, Pan H (2023) Status and trends of international cooperation in Brazil's new energy sector. Int Pet Econ 31(5):48–55, 62
11. Wang M, Chen Y (eds) (2021) Series on climate change economics: global climate governance. China Social Sciences Press
12. Rodrigues M (2023) How is Brazil's President Lula doing on climate? Experts rate his performance. Nature. https://www.nature.com/articles/d41586-023-02511-x. Accessed 22 Oct 2023
13. Wang H, Liu B (2024) The impact of Brazil's climate policy actions on global climate governance. Journal Title [replace with actual journal name]
14. Liu M (2022) Brazil's shifting stance on global climate governance norms and its roots: a policy autonomy perspective. Doctoral dissertation, University of International Business and Economics
15. Nan B (2024) A "major victory" for Lula one year after taking office: Brazil's rainforest deforestation area decreased in 2023. The Paper. https://www.thepaper.cn/newsDetail_forward_26017156. Accessed 21 Jan 2024
16. Zhao B (2017) Grouping: the path choice for emerging powers to participate in global climate governance. Int Forum 19(2):8–15, 79

Chapter 9
Climate Governance in Africa: A Continent's Responses to Global Challenges

The African continent stands at the epicenter of the global climate crisis, embodying its most profound paradoxes and pressing urgencies. While contributing minimally to historical greenhouse gas emissions, its nations face disproportionate vulnerabilities, where climate impacts directly threaten hard-won development gains and regional stability. This chapter examines the multifaceted landscape of African climate governance over the past three decades, moving from the continental to the national level to reveal a complex tapestry of ambition, constraint, and strategic adaptation. Through the lens of the African Union as a regional integrator, South Africa as an industrializing power navigating a just transition, and Qatar as a hydrocarbon-rich state balancing immense wealth with existential risk, we explore the diverse pathways, shared challenges, and evolving agency of African actors in the global fight against climate change. Their collective journeys underscore a critical shift from being passive victims of a changing climate to becoming assertive architects of their own sustainable futures.

9.1 The African Union as a Regional Integrator and Advocate

Against the backdrop of insufficient global momentum for climate cooperation and increasingly prominent climate security threats, regional international organizations are playing an ever more critical role in shaping the climate security governance agenda. The African Union (AU), as the largest and most representative regional organization in Africa, has, over the past three decades, undergone a climate governance journey that profoundly reflects the continent's exploration, challenges, and ambitions in confronting global climate change.

X. Zou, *Thirty Years of Global Climate Governance*, Energy, Climate Change & Sustainability in the Global South, https://doi.org/10.1007/978-981-95-8457-4_9

9.1.1 Drivers and Context of AU Climate Governance

Africa ranks among the region's most severely impacted by climate change. The Intergovernmental Panel on Climate Change (IPCC) Sixth Assessment Report further substantiates, from a scientific perspective, the multi-dimensional threats posed by climate change to the African region. Dominated by tropical climate zones with vast arid areas, including the Sahara Desert spanning over 9 million square kilometers, the continent is particularly vulnerable. With an estimated population exceeding 1.4 billion, growing at an annual rate of 2.3%, and a significant proportion living in poverty, the socio-economic impacts are magnified. According to projections by the African Climate Policy Centre of the UN Economic Commission for Africa, climate-related losses and damages in Africa between 2020 and 2030 are estimated to range between USD 290 billion and USD 440 billion. [1]

Crucially, while Africa contributes less than 10% of global carbon emissions, it remains the world's most climate-vulnerable region. The World Meteorological Organization's "State of the Climate in Africa 2022" report explicitly states that despite Africa's minor contribution to global greenhouse gas (GHG) emissions, it suffers disproportionately from severe climate impacts. These manifest as compromised food security, ecosystem degradation, hindered socio-economic development, exacerbated displacement and migration, and an increased threat of conflicts over dwindling resources. [2] In 2022 alone, over 110 million people across the continent were directly affected by weather-, climate-, and water-related hazards, resulting in economic losses exceeding USD 8.5 billion. [2]

9.1.2 Institutionalization and Strategic Evolution of AU Climate Governance

The AU's climate governance process exhibits distinct characteristics of institutionalization and strategic planning. A significant milestone was reached in 2007 when the 8th AU Assembly of Heads of State and Government elevated climate change to a key agenda item. Concurrently, the AU launched the "Great Green Wall for the Sahara and Sahel Initiative," marking a notable elevation of climate issues within the Union's priorities. [3] In 2011, the AU Assembly authorized the Climate for Development in Africa (ClimDev-Africa) Programme, aimed at laying a solid foundation for Africa's climate response and fostering close cooperation with relevant partners.

A pivotal step came in 2014 with the drafting of the *African Union Strategy on Climate Change*. This strategy pioneered policy innovation, particularly in advancing climate adaptation, and established a necessary institutional platform for solidarity and collaboration among African nations on this issue. [4] The subsequent launch of *Agenda 2063: The Africa We Want* in 2015 formally identified enhancing the capacity to address climate change and achieve sustainable development as a critical goal for the continent's development. Consequently, climate action became one of the most frequently cited keywords in Africa's future development strategy.

Entering the 2020s, the AU's climate governance agenda deepened further. The 2022 *African Common Position on Energy Access and Just Transition* outlined short-, medium-, and long-term energy development pathways, proposing an energy transition that simultaneously meets development needs and anticipating the establishment of a new global climate finance architecture by 2025. The year 2023 proved significant. The UNFCCC COP27, held in Egypt, was widely termed the "African COP," with its core focus squarely on the unfulfilled climate finance commitments of developed nations. The same year witnessed the African Climate Summit adopting the *Nairobi Declaration*, which called for global collaboration to reduce GHG emissions and urged developed countries to deliver on their pledges of funding and technological assistance. Furthermore, the AU's accession to the G20 summit in 2023 positioned it to play a vital role in global climate governance, amplifying a strong African voice on the international stage.

9.1.3 Multidimensional Challenges and Constraints

The AU faces a complex array of internal and external challenges in advancing its climate governance agenda.

Firstly, Western international climate hegemony constrains the AU's efforts. The implementation of the European Union's Carbon Border Adjustment Mechanism (CBAM) has been criticized for severely undermining the most-favoured-nation treatment enjoyed by African member states under the Cotonou Agreement for exports to Europe. This mechanism is argued to effectively negate the principle of "Common but Differentiated Responsibilities" established by the UNFCCC. [5]

Secondly, the AU grapples with a lack of economic and technological capacity. Estimates from the *Africa's Adaptation Gap Report II* suggested that the cost of science and technology for climate adaptation in Africa in 2020 was between USD 7 billion and USD 15 billion, a figure beyond the reach of many least-developed countries on the continent. [6] This dual scarcity of funding and technology severely curtails the realization of the AU's climate ambitions.

Moreover, climate issues in Africa are deeply intertwined with numerous social development challenges, creating a more complex scenario. Climate disasters often trigger secondary crises such as plagues and famines, which can easily escalate into large-scale humanitarian emergencies and foster the growth of extremist groups. The compounding effects of climate-induced food security crises, poverty, and internal "climate migration" (the World Bank predicts up to 86 million internal climate migrants in Sub-Saharan Africa by 2050) constitute an intricate governance dilemma.

Finally, coordinating the relationship between the AU and its member states remains a fundamental challenge. Significant disparities in resource endowments, development levels, and economic structures among member states lead to divergent manifestations of climate risks, varying levels of vulnerability, and differing governance capacities. These differences, in turn, foster profound divisions in national interests and positions regarding climate policy. For instance, Nigeria and South Africa, due to their heavy economic reliance on fossil fuels, often adopt cautious stances on energy transition policies. Simultaneously, although the *Constitutive Act of the African Union*赋予 (confers) the AU with the function of intergovernmental coordination, its policy coordination and guidance role remains factually limited as it lacks compulsory jurisdiction over member states. [7, 8]

9.1.4 Future Prospects and Pathways

Looking ahead, the prospects for AU climate governance will likely unfold across several core dimensions:

In the energy sector, the AU is set to delineate short-, medium-, and long-term energy development pathways. Accelerating the deployment of clean energy sources like solar, wind, and hydropower is crucial to meet the continent's rapidly growing electricity demand and ultimately achieve universal electricity access. Regarding the implementation of Nationally Determined Contributions (NDCs), the AU will leverage policy analysis to assist African policymakers in formulating climate response strategies that promote sustainable development. By providing capacity-building support to member states, the AU aims to foster the effective utilization of natural resources and develop climate-resilient and inclusive economies.

Resilience building is increasingly viewed as a core pathway for the AU to address climate security challenges. The AU Peace and Security Council has emphasized the need for comprehensive access to climate security risk information, accelerated completion of relevant risk assessments, and the development of reliable early warning systems based on these assessments. [9] The Africa Multi-Hazard Early Warning and Early Action System initiative, launched by the AU Commission, represents a significant advancement in this area. The AU's climate strategy is committed to comprehensively enhancing the climate resilience of key sectors—including agriculture and food systems, water resources, energy, infrastructure, and transport—thereby driving Africa's green transition.

To achieve these objectives, the AU must further facilitate dialogue among member states, harmonize their positions, and strengthen interactions with international mechanisms like the United Nations. Integrating climate security issues into various continental transition plans and promoting synergy between the AU's *Agenda 2063* and the UN's *2030 Agenda for Sustainable Development* will be essential.

9.2 South Africa's Path from Periphery to Center

South Africa, as the continent's most industrialized nation and a major emitter, presents a compelling case study. Its three-decade journey in climate governance reflects not only a complex domestic balancing act between environment, development, and justice but also its evolving identity and status in international climate politics—transforming from a state marginalized by apartheid into an indispensable and active participant, even a leader, in global climate governance.

9.2.1 Drivers of South Africa's Climate Engagement

South Africa's commitment to climate governance is shaped by a confluence of natural imperatives and political strategy. Environmentally, the country faces acute vulnerabilities. Characterized by a predominantly arid climate with highly uneven water distribution—over sixty percent of runoff concentrated in just twenty percent of the territory—it is exceptionally susceptible to climate shifts. [10] Climate change exacerbates rainfall unpredictability, increases high-temperature evaporation, and intensifies extreme weather events, posing severe threats to water security, agricultural yields (especially staple crops like maize), and human health (e.g., expanding ranges of vector-borne diseases such as malaria). South Africa's 2010 National Climate Change Response Green Paper projected temperature increases of 3–4 °C in coastal regions and approximately 6–7 °C inland by around 2050, portending significant damage to biodiversity and reduced vegetation cover. [11]

Structurally, South Africa's energy system is a core driver of its climate policy. As the world's seventh-largest coal producer, coal accounts for about 70% of its primary energy consumption. This profound dependency makes it Africa's largest GHG emitter. Consequently, the majority of its emissions stem from coal combustion for energy and industrial processes. Its emissions trajectory displays a "dual character": total volumes, while generally trending upwards, are significantly higher than those of fellow developing economies like India and Brazil, yet remain lower than many developed countries.

Politically, the establishment of the "new South Africa" in 1994 provided a fresh foundation for engaging in global affairs. Climate governance, as a universally relevant and non-ideological global issue, offered an ideal platform for the nation to shed the negative international legacy of apartheid, project its new democratic identity, and secure international standing and influence.

9.2.2 Historical Evolution and Policy Deepening

South Africa's climate governance can be delineated into five distinct, evolving phases, marking a transition from passive participation to proactive leadership.

Phase 1: Passive Participation (1992–1994)

During this period, domestic focus remained squarely on resolving the apartheid conflict, and climate governance was not systematically addressed. The primary outcome was following the international community in the development and signing of the United Nations Framework Convention on Climate Change (UNFCCC). The establishment of the National Climate Change Committee (NCCC) in 1994 is widely regarded as the symbolic starting point of South Africa's formal climate politics. [12]

Phase 2: Domestic Foundation-Laying (1994–2002)

Post-apartheid, South Africa actively built its domestic policy and legal infrastructure. The new 1996 Constitution enshrined environmental protection as a citizen's right for the first time. The National Environmental Management Act of 1998 was promulgated to enhance coordination between environmental authorities and other government departments. [13] South Africa ratified the UNFCCC in 1997 and acceded to the Kyoto Protocol in 2002. The same year, it hosted the World Summit on Sustainable Development in Johannesburg, with then-President Thabo Mbeki serving as conference chair—an event that significantly elevated climate change to the highest level of the national political agenda.

Phase 3: Active Engagement (2002–2009)

This phase saw South Africa strengthening both its domestic and international roles. Domestically, it released its first National Climate Change Response Strategy in 2004. In 2005, climate change was formally tabled as a policy issue at a dedicated government conference (the Midlands Climate Action Meeting). The "Midlands Plan" emerging from this meeting directly contributed to the development of the "Long-Term Mitigation Scenarios" (LTMS) in 2008, which proposed peaking national emissions between 2020–2025, followed by a transition to a low-carbon economy. [14] Internationally, the 2002 Johannesburg Summit showcased South Africa's unique ability to mediate conflicts and foster cooperation in multilateral negotiations, heralding the beginning of its environmental diplomacy. At the 2007 Bali Climate Conference (COP13), South Africa, operating within the "G77 and China" framework, acted as a key negotiator, actively pursuing a central role in global climate governance discussions. [15]

Phase 4: The Quest for Leadership (2009–2015)

This period was pivotal for establishing South Africa's influence on the international climate stage. In 2009, it joined with China, Brazil, and India to form the "BASIC" group, creating an effective coordination mechanism for unified advocacy. At the Copenhagen Climate Conference (COP15), President Jacob Zuma announced quantified emission reduction targets: 34% by 2020 and 42% by 2025, relative to a business-as-usual trajectory. [16] The pinnacle of this phase was South Africa's hosting of the Durban Climate Conference (COP17) in 2011. When negotiations reached an impasse, South African diplomacy demonstrated formidable mediation skills, extending the conference by two days to broker the "Durban Package." This outcome was crucial as it launched the process that eventually led to the Paris Agreement, signifying South Africa's transformation from a learner in global climate governance to a leading, influential actor.

Phase 5: Post-Paris Planning and Implementation (2015-Present)

Aligning with the Paris Agreement framework, South Africa embarked on systematic planning for carbon neutrality. Key measures included implementing a Carbon Budget system in 2016; passing the Carbon Tax Act in 2019, making it the first African nation to institute a carbon tax; and enacting the Climate Change Bill in 2018 (awaiting final signature as of knowledge cutoff) to address coordination challenges across different tiers of government. In February 2020, South Africa submitted its "Low Emission Development Strategy 2050" to the UN, outlining a "Peak-Plateau-Decline" emissions pathway and setting NDC targets for 2025 (398–510 Mt CO2-eq) and 2030 (350–420 Mt CO2-eq). In 2021, an updated NDC solidified its commitment to achieve net-zero carbon emissions by 2050.

9.2.3 Core Challenges and Future Orientation

Despite notable achievements, South Africa's climate governance faces stern tests. Firstly, inter-departmental coordination remains a hurdle. The Department of Forestry, Fisheries and the Environment (DFFE), as the lead agency, has limited capacity and authority to steer the entire government effectively. Overlapping mandates with powerful entities like the National Treasury and various inter-ministerial committees often lead to fragmented policies and inconsistent implementation. Secondly, financial constraints are severe. While international support mechanisms like the Just Energy Transition Partnership (JETP), which pledges $8.5 billion, are welcome, the country faces a massive financing gap estimated at ZAR 660 billion. Its deep dependence on coal implies exorbitant costs for phasing out coal-fired power, creating a heavy reliance on international aid that has often been slow to materialize. Thirdly, deficits in technical capacity and reliable data further impede effective governance.

Looking forward, South Africa's climate trajectory necessitates focused efforts on several fronts. First, enhancing stakeholder engagement and inter-departmental collaboration is critical, potentially through high-level coordinating mechanisms that institutionalize climate action beyond individual interactions. Second, developing a comprehensive financial strategy and establishing dedicated institutional capacity for climate adaptation are essential. This should be coupled with efforts to diversify financing channels, leveraging both domestic green fund initiatives and international cooperation platforms (e.g., BASIC, BRICS, South-South Cooperation). South Africa stands at a crossroads on its path "from periphery to center." Its success will depend on its ability to forge robust domestic consensus and, in concert with other emerging economies, continue to shape a global climate governance regime that is both ambitious and equitable, thereby accelerating its own domestic green transition and low-carbon development.

9.3 Qatar's Dilemma and Strategic Pivot as a Hydrocarbon Giant

Qatar, a quintessential Gulf state, presents a critical case study of a resource-rich economy navigating the global climate imperative. As one of the world's highest per-capita carbon emitters and the largest producer and exporter of liquefied natural gas (LNG), its climate governance path vividly illustrates the structural contradictions, transition pressures, and strategic choices confronting fossil fuel-dependent nations.

9.3.1 Acute Challenges and Inherent Vulnerabilities

Qatar's climate challenges are intrinsically linked to its natural geography and economic structure. The World Meteorological Organization confirming 2023 as the hottest year on record underscores the global threat, with glacier melt and sea-level rise posing a direct risk to 97% of Qatar's population, which is concentrated in coastal urban areas. A recent study by the Center for Global Development places Qatar among the top 20 countries most threatened by sea-level rise by 2050.

Economically, Qatar exhibits one of the world's highest dependencies on a hydrocarbon-based economy. Oil and gas exports constituted 43.4% of its GDP in 2022, contributing over USD 104 billion. [17] However, its energy transition faces significant structural bottlenecks. Qatar's small land area (11,521 km^2) physically constrains the large-scale deployment of renewable energy infrastructures like solar and wind farms. Data revealing that renewable energy accounted for a mere 0.0001% of its total electricity consumption from 2016 to 2020, starkly contrasted with the global average of 18.15% during the same period, highlights the immense scale of this challenge. [18] In an era of accelerating global energy transition, reconciling the short-term economic rents from fossil fuel exports with the long-term imperative of a low-carbon future constitutes the central dilemma for Qatar's sustainable development.

9.3.2 Strategic Deployment and Practical Explorations

In response, Qatar has progressively constructed a climate governance framework through institutional reforms, policy planning, technological investment, and international engagement.

At the institutional and strategic level, a significant move was the formal establishment of the Ministry of Environment and Climate Change in 2021, tasked with formulating public policies for environmental protection and emission reduction. That same year, Qatar launched its National Climate Change Plan, committing to a 25% reduction in greenhouse gas emissions by 2030 and aspiring to achieve net-zero emissions before 2050. The plan prioritizes reducing air pollutants, developing green and clean energy, and restoring biodiversity. This climate agenda is strategically aligned with Qatar's National Vision 2030, which sets goals for developing a sustainable society, promoting economic and industrial diversification, and harnessing multiple energy sources, supported by 35 specific measures and over 300 targeted initiatives to bolster national climate resilience. [19]

In technology application and energy transition, Carbon Capture, Utilization, and Storage (CCUS) is deemed a cornerstone solution. The Qatar Carbonates and Carbon Storage Research Centre, a collaboration between QatarEnergy, Shell, and Qatar Science & Technology Park, has been instrumental in investing heavily in CO_2 capture technology. QatarEnergy's LNG subsidiary already captures 2.1 million tonnes of CO_2 annually from the Ras Laffan liquefaction plant, with plans to increase this capacity to 5 million tonnes per year by 2025. [20] In renewables, the commissioning of the Al Kharsaah Solar Power Plant in 2022 marked a pivotal milestone. This facility meets 10% of Qatar's peak energy demand and is projected to prevent 26 million tonnes of CO_2 emissions over its operational lifetime. [21]

On the international stage, Qatar has adeptly leveraged mega-events to project its climate commitments. The 2022 FIFA World Cup served as a concentrated showcase of its climate governance practices. The nation pledged to deliver a "carbon-neutral" tournament and committed to purchasing 1.8 million carbon offset credits from the Global Carbon Council (GCC). According to FIFA's accounting report, the total GHG emissions for the World Cup were approximately 3.63 million tonnes of CO_2-equivalent. [22] While this approach attracted criticism from some environmental groups alleging "greenwashing," it undeniably accelerated domestic practices in sustainable construction, renewable energy application, and carbon management. In climate finance, Qatar pledged USD 100 million during the 2019 UN General Assembly and has set an ambition to mobilize a cumulative USD 100 billion between 2020 and 2025 to support Small Island Developing States and Least Developed Countries in coping with climate change.

9.3.3 *Inherent Contradictions and Future Prospects*

Qatar's climate strategy is characterized by a palpable "duality." On one hand, it demonstrates a willingness to transition towards a greener economy by setting net-zero targets, investing in solar power, and pioneering CCUS. On the other hand, its economic foundation remains deeply anchored in fossil fuel exports, with plans to significantly increase LNG production capacity from 77 to 110 million tonnes per annum. This strategic contradiction is evident in the stark contrast between its National Vision 2030 target of sourcing 20% of its electricity from renewables and the current near-zero baseline.

The efficacy of Qatar's future climate governance will hinge on several critical factors. First is its ability to genuinely overcome technological and natural constraints to deploy renewable energy at a scale sufficient to reduce domestic reliance on natural gas for power generation. Second is the capacity to effectively leverage revenues from its traditional energy sector to fund a green transition and achieve meaningful economic diversification, thereby escaping the "resource curse." Third, within international climate negotiations, Qatar must navigate a delicate balance: defending its legitimate interests as a major energy exporter while proactively integrating into the global climate governance system to enhance the credibility and transparency of its commitments.

Notwithstanding the significant headwinds—including technological bottlenecks, profound path dependency, and volatile geopolitics impacting energy prices—the global net-zero consensus, reinforced through forums like COP28, injects both fresh impetus and urgency into the green development pathways of Qatar and its regional counterparts.

9.4 Concluding Remarks

In conclusion, the trajectories of the African Union, South Africa, and Qatar illuminate the diverse yet interconnected realities of climate governance in an African context. The AU's struggle to forge a unified continental front highlights the perpetual challenge of coordinating disparate national interests under the shadow of inadequate international finance and technological transfer.

South Africa's journey "from periphery to center" demonstrates the immense political and economic costs of decarbonizing a coal-dependent economy, underscoring the non-negotiable role of international support for a just transition. Qatar's case lays bare the fundamental tension at the heart of the global energy transition, where even the wealthiest fossil fuel producers are compelled to strategize for a post-carbon world, yet remain tethered to the very industries that define their present. Together, these narratives affirm that effective climate governance in Africa and beyond is not merely a technical or environmental issue, but a deeply political one, requiring nuanced strategies that reconcile the imperatives of equity, development, and survival on an increasingly volatile planet. Their successes and struggles offer indispensable lessons for a global community still grappling with how to operationalize climate justice and shared responsibility.

References

1. African Climate Policy Centre, UN Economic Commission for Africa (n.d.) Climate change and development in Africa. https://www.uneca.org/cr4d. Accessed 10 Oct 2024
2. World Meteorological Organization (2022) State of the climate in Africa 2022. WMO-No. 1300, Geneva
3. African Union (2007) Assembly decision on climate change and development in Africa. Assembly/AU/Dec.156 (VIII)
4. African Union (2021) Draft African Union strategy on climate change (AMCEN-15-REF-11). https://www.ucis.pitt.edu/africa/sites/default/files/Downloads/PDFs/AU%20Strategy%20on%20Climate%20Change.pdf. Accessed 21 Oct 2023
5. Deng Y (2022) Addressing climate change: Africa's exploration and the significance of China-Africa cooperation. J Univ Chin Acad Soc Sci 42(10):106–121, 143–144. (Note: Translated title and source; original Chinese citation provided by user).
6. United Nations Environment Programme (2015) Africa's adaptation gap 2: bridging the gap – mobilising sources. https://www.unep.org/resources/report/africas-adaptation-gap-2-bridging-gap-mobilising-sources. Accessed 21 Oct 2023
7. Wu K, Jin L (2022) Achievements and problems of the African Union's public health governance as seen from the response to COVID-19. West Asia and Africa, (6), 39. Translated title and source; original Chinese citation provided by user, Note
8. Medinilla A (2021) AU-EU cooperation on climate and energy: in search of a common narrative. ECDPM Briefing Note No. 133. https://sloga-platform.org/wp-content/uploads/2021/05/AU-EU-Cooperation-Climate-…. Accessed 21 Oct 2023. (Incomplete URL from source, retained as per original)
9. African Union Peace and Security Council (2021) Communique. PSC/PR/COMM.1051 (2021). 26 Nov 2021
10. Ilemobade AA, Adewumi JR, van Zyl JE (2012) Framework for assessing the viability of implementing dual water reticulation systems in South Africa. Water SA 35(2):216–227
11. The Government of the Republic of South Africa (2010) National climate change response green paper
12. Zhao B (2022) From periphery to center: an analysis of the development of South Africa's climate governance. West Asia and Africa, (1). Translated title and source; original Chinese citation provided by user, Note
13. Zhang X, Zhang N (2022) Environmental Protection legal systems and risk response in southern Africa. African Studies, (12). Translated title and source; original Chinese citation provided by user, Note
14. Hallding K et al (2011) Together Alone: Basic countries and the climate change conundrum. Nordic Council of Ministers, Copenhagen, p 50
15. Zhao B (2022) Op. Cit
16. Liu Q (2009) South Africa announces conditional emission reduction targets: 34% reduction by 2020. Phoenix News. https://news.ifeng.com/world/special/gebenhagenqihou/zuixin/200912/1208_8755_1465837.shtml. Accessed 2 Mar 2026. (Note: Translated title and source from Chinese)
17. Planning and Statistics Authority, State of Qatar (2023) Report on quarterly estimates of gross domestic product by economic activity and components of expenditure for Qatar economy. https://www.psa.gov.qa/en/statistics1/Pages/LatestStats/28112023.asp. Accessed 2 Mar 2026

18. Planning and Statistics Authority, State of Qatar (2022) Affordable and clean energy, sustainable development goals. https://gis.psa.gov.qa/GISApps/SDG/Data/PDF/En/SDG7.pdf. Accessed 2 Mar 2026
19. The Peninsula (2021) Qatar unveils environment and climate change strategy. https://thepeninsulaqatar.com/article/29/10/2021/qatar-unveils-environment-and-climate-change-strategy. Accessed 2 Mar 2026
20. Global CCS Institute (2023) Global status of CCS 2023: executive summary. https://www.globalccsinstitute.com/resources/publications-reports-research/global-status-of-ccs-2023-executive-summary/. Accessed 2 Mar 2026
21. Power Technology (2023) Al Kharsaah solar power project. https://www.power-technology.com/projects/al-kharsaah-solar-power-project/. Accessed 2 Mar 2026
22. FIFA (2022) Greenhouse gas accounting report: FIFA World Cup 2022™. https://visitqatar.com/content/dam/visitqatar/img/things-to-do/get-inspired/fifa-files/greenhouse-gas-accounting-report.pdf. Accessed 2 Mar 2026

Chapter 10
ASEAN's Role in Climate Governance: Review Challenges, and Future Prospects

This chapter focuses on the background and evolutionary process of ASEAN's climate governance, with particular attention to its dual challenges at present: at the international level, the lack of a legally binding mechanism and the susceptibility to great-power rivalry; at the national level, disparities in development levels among member states result in uneven governance practices and limited regional coordination. Based on this analysis, the paper further proposes a set of policy recommendations, including reforming the "ASEAN Way" decision-making mechanism, actively responding to the influence of global geopolitical changes, formulating differentiated governance targets according to national development capacities, and strengthening both intra-regional and extra-regional cooperation.

10.1 Introduction

The Association of Southeast Asian Nations (ASEAN), established in 1967 amid the geopolitical tensions of the Cold War, has evolved into a pivotal regional organization representing over 660 million people across ten member states: Brunei, Cambodia, Indonesia, Laos, Malaysia, Myanmar, the Philippines, Singapore, Thailand, and Vietnam. Over the past three decades, from the 1992 Rio Earth Summit to the 2021 Glasgow Climate Conference (COP26), ASEAN's engagement with global climate governance has transitioned from peripheral participation to a more structured, albeit uneven, regional framework. This chapter examines ASEAN's climate governance trajectory, highlighting its historical development, persistent challenges, and prospective strategies. Drawing on the principles of the United Nations Framework Convention on Climate Change (UNFCCC) and subsequent agreements like the Kyoto Protocol and Paris Agreement, it underscores ASEAN's unique position as a bridge between developing and developed worlds in addressing climate change.

X. Zou, *Thirty Years of Global Climate Governance*, Energy, Climate Change & Sustainability in the Global South, https://doi.org/10.1007/978-981-95-8457-4_10

Climate change poses an existential threat to Southeast Asia, a region characterized by its vulnerability to sea-level rise, extreme weather events, and biodiversity loss. ASEAN's collective response has been shaped by shared geographical proximities, economic interdependencies, and political histories of colonialism and non-interference. However, the "ASEAN Way"—emphasizing consensus, sovereignty, and non-intervention—has both facilitated cooperation and constrained binding commitments. This chapter argues that while ASEAN has made strides in integrating climate issues into its socio-cultural and economic pillars, structural limitations, including developmental disparities and geopolitical entanglements, continue to impede effective governance. By adopting a multi-level governance lens, which incorporates international, regional, national, and subnational actors, ASEAN can enhance its resilience and contribute more robustly to global efforts.

The analysis begins with the foundational background of ASEAN's climate engagement, followed by its evolutionary phases. It then delineates current challenges at international and national levels, proposes responsive strategies, and explores multi-level governance as a transformative approach. Case studies on Indonesia and Thailand illustrate national variations, while a discussion of international collaborations highlights external synergies. Ultimately, this chapter posits that ASEAN's future in climate governance hinges on balancing internal cohesion with adaptive reforms, ensuring equitable and sustainable development amid escalating global risks [1].

10.2 Background of ASEAN Climate Governance

ASEAN's climate governance is rooted in the region's interconnected geographical, economic, and political contexts, which simultaneously foster cooperation and expose vulnerabilities. Member states share common interests in addressing climate change, yet their diverse capacities necessitate tailored approaches. This section elucidates these foundational elements, providing a basis for understanding ASEAN's evolving strategies.

10.2.1 Geographical Environment

Southeast Asia's geography renders it one of the world's most climate-vulnerable regions. Predominantly comprising island, peninsular, and maritime nations, ASEAN countries are clustered around the equator, sharing tropical climates, extensive coastlines, and rich biodiversity hotspots such as coral reefs and rainforests. This proximity facilitates transboundary environmental challenges: shared river basins like the Mekong Delta and common marine ecosystems in the South China Sea amplify the spillover effects of climate impacts [2]. With a total population exceeding 660 million—approximately 8.5% of the global total—about 80% reside within 100 km of coastlines, heightening exposure to sea-level rise [3]. Projections from the Asian Development Bank (ADB) indicate that unchecked greenhouse gas emissions could elevate regional temperatures by 4.8 °C and sea levels by 70 cm by 2100, displacing millions from low-lying areas in countries like Vietnam and the Philippines [4]. Freshwater scarcity, exacerbated by warming, threatens over 20 million people, while tectonic activity in the "Ring of Fire" intensifies disasters such as typhoons, floods, and earthquakes [5].

These shared vulnerabilities underscore ASEAN's interdependence: climate change does not respect borders, demanding collective action. For instance, deforestation in Indonesia contributes to regional haze pollution, affecting air quality across borders [6]. Thus, geographical factors not only heighten risks but also lay the groundwork for collaborative mechanisms, such as the ASEAN Agreement on Transboundary Haze Pollution (2002), which exemplifies early regional responses [7].

10.2.2 Economic Linkages

ASEAN's economies, predominantly developing with a combined GDP of over $3 trillion, are intricately linked through trade, resource dependencies, and shared growth aspirations [8]. Many rely on climate-sensitive sectors: agriculture employs 30–40% of the workforce in countries like Cambodia and Laos, while tourism in Thailand and Indonesia hinges on pristine beaches and ecosystems [9]. Complementary trade patterns—e.g., Indonesia and Malaysia exporting oil to importers like Thailand and the Philippines—bolster regional integration, with Singapore serving as a financial and logistical hub [10].

However, climate change inflicts disproportionate economic damages, estimated by the International Monetary Fund (IMF) to exceed global averages due to water shortages, food insecurity, and health risks [11]. Agricultural yields could decline by 20–30% in rice-producing nations, threatening food security for 200 million people [12]. Tourism, generating $300 billion annually, faces erosion from coastal degradation [13]. These impacts propel ASEAN toward climate governance as an economic imperative, evident in initiatives like the ASEAN Economic Community's emphasis on sustainable growth [14]. Economic linkages also reveal disparities: wealthier members like Singapore invest in green technologies, while poorer ones like Myanmar struggle with adaptation funding [15]. This asymmetry necessitates differentiated strategies, aligning with global principles like common but differentiated responsibilities (CBDR) [16].

10.2.3 Political Factors

ASEAN's political landscape, marked by post-colonial legacies and a commitment to sovereignty, profoundly influences its climate governance. Many members endured colonial rule, fostering a peripheral role in global politics until regional integration elevated their status [17]. The 1976 Treaty of Amity and Cooperation reinforced non-interference, enabling cooperation on shared interests without compromising autonomy [18]. Climate issues initially emerged as environmental concerns but evolved into political priorities amid global agendas. The "ASEAN Way" promotes consensus, avoiding coercive measures, which suits diverse regimes—from democracies like Indonesia to authoritarian systems like Vietnam [19]. However, this approach limits binding commitments, contrasting with more hierarchical models like the EU's [20]. Politically, ASEAN navigates great-power rivalries: U.S. "Pivot to Asia" and China's Belt and Road Initiative (BRI) intersect with climate diplomacy, where cooperation risks instrumentalization [21]. Historical wariness of external influence complicates partnerships, yet ASEAN's centrality in forums like the East Asia Summit amplifies its voice [22]. Overall, political factors provide a resilient foundation for climate collaboration, albeit one tempered by sovereignty concerns.

10.3 Evolution of ASEAN Climate Governance

ASEAN's approach to climate governance has undergone significant transformation over the past three decades, reflecting both global trends and regional priorities. A pivotal shift occurred around 2007, marking a transition from sporadic, exploratory efforts to more institutionalized and proactive measures. This evolution can be delineated into two phases: pre-2007, characterized by initial agenda inclusion, and post-2007, marked by deepening frameworks and multi-stakeholder engagement. The demarcation aligns with key global milestones, such as the Kyoto Protocol's entry into force in 2005 and the surge in international climate discourse in 2007, which prompted ASEAN to elevate climate issues to summit-level discussions [23].

10.3.1 Preliminary Exploration and Agenda Inclusion (Pre-2007)

Prior to 2007, ASEAN's engagement with climate change was tentative, emerging primarily in response to international developments rather than endogenous initiatives. The establishment of the Intergovernmental Panel on Climate Change (IPCC) in 1988 and the 1992 United Nations Conference on Environment and Development (UNCED) in Rio de Janeiro introduced climate change to the global political agenda, prompting ASEAN's initial responses. At the 1992 ASEAN Ministerial Meeting on the Environment (AMME), members acknowledged the need for urgent measures, aligning with the UNFCCC's adoption that year [24].

ASEAN countries actively participated in early UNFCCC processes, with most ratifying the Convention by the mid-1990s. The Kyoto Protocol, adopted in 1997, received positive responses from ASEAN states, which successively ratified it, contributing to its 2005 enforcement. This period saw climate issues framed as environmental extensions rather than core political concerns. The 1997 ASEAN Vision 2020 articulated aspirations for a "clean and green ASEAN," emphasizing regional mechanisms for environmental protection and sustainable resource use [25]. To operationalize this, the 2003 Bali Concord II launched the ASEAN Community, integrating climate cooperation into the Socio-Cultural Community pillar under AMME oversight [26]. Key events, such as the 2007 UNFCCC Conference of the Parties (COP13) in Bali, hosted by Indonesia, catalyzed ASEAN's integration of climate into regional agendas. Pre-2007 efforts were largely declarative, focusing on capacity building and information sharing, with limited binding commitments. For instance, early collaborations with the United Nations Environment Programme (UNEP) emphasized vulnerability assessments, but lacked robust enforcement [27]. This phase laid foundational norms, such as people-centered development, but highlighted the need for deeper institutionalization amid rising global pressures.

10.3.2 Deepening and Institutionalization (Post-2007)

Post-2007, ASEAN's climate governance accelerated, driven by global momentum from forums like the World Economic Forum and EU summits, which elevated climate to high politics. ASEAN responded by upgrading discussions to summit levels, issuing specialized declarations, and proliferating policy instruments. This era is characterized by three core features: integration with sustainable development, multi-sectoral frameworks, and multi-stakeholder participation [28].

First, climate change was inextricably linked to sustainable development, viewing it as a barrier to economic progress. ASEAN promoted low-carbon transitions through renewable energy, energy efficiency, and clean technologies. National plans proliferated: Cambodia's Sustainable Electricity Action Plan (2002–2012), Malaysia's National Sustainable Energy Policy (2010) and Sustainable Energy Act (2011), and Indonesia's Energy and Electricity Laws institutionalized these goals [29]. Regionally, the ASEAN Plan of Action for Energy Cooperation (APAEC) 2016–2025 targeted 23% renewable energy by 2025, reflecting a shift toward green growth [30].

Second, ASEAN constructed multi-sectoral governance frameworks, conceptualizing climate as a multidimensional security issue. Policies spanned environment, agriculture, forestry, disaster management, health, and energy security. The 2014 ASEAN Joint Statement on Climate Change emphasized "people-centered" approaches, addressing low-carbon societies, energy access, food security, and livelihoods [31]. Mechanisms like the ASEAN Working Group on Climate Change

(AWGCC), established in 2009, coordinated these efforts, ensuring synergy across pillars [32].

Third, governance expanded beyond governments to include private sectors, NGOs, and communities. ASEAN encouraged diverse participation in risk mitigation, fostering networks with civil society. For example, forums like the ASEAN Civil Society Conference facilitate dialogue, while partnerships with international NGOs support community-based adaptation [33]. This inclusivity aligns with global trends, such as the Paris Agreement's emphasis on non-state actors.

Overall, post-2007 institutionalization has enhanced ASEAN's climate resilience, though implementation varies. From 2007 to 2023, over 50 climate-related documents were issued, compared to fewer than 10 pre-2007, demonstrating maturation [34]. Yet, as global temperatures breached 1.5 °C thresholds in 2024, ASEAN's progress underscores the urgency for further ambition [35].

10.4 Current Challenges

Despite advancements, ASEAN's climate governance faces multifaceted challenges at international and national levels. These stem from institutional limitations, geopolitical dynamics, developmental disparities, and coordination deficits, impeding collective efficacy.

10.4.1 International Level Challenges

At the international level, ASEAN grapples with the "ASEAN Way's" inherent constraints and escalating uncertainties in global cooperation.

The "ASEAN Way," prioritizing consensus and non-interference, originated in traditional security contexts to safeguard sovereignty [36]. In climate governance, however, it manifests as declarative texts lacking operationality and enforceability. Absent monitoring or sanctions, cooperation remains "soft," yielding suboptimal outcomes [37]. Comparatively, the EU's Fit for 55 package, enshrined in the European Climate Law, mandates a 55% emissions reduction by 2030 from 1990 levels, with binding national targets based on per-capita GDP [38]. Reforms to the EU Emissions Trading System (ETS) enforce a 61% cut by 2030, balancing costs via adjustments [39].

In contrast, ASEAN's plans offer directional guidance without unified, binding obligations. The 2021 ASEAN State of Climate Change Report advocates collaboration but highlights divergences: Indonesia, emitting over 60% from coal, pledged net-zero by 2060 contingent on finance, while its 2020 Omnibus Law relaxed environmental rules, sparking controversy [40]. Singapore, implementing carbon pricing, advances via industrial transformation and international ties, serving as a benchmark [41]. Without enforcement, member compliance varies, limiting regional impact.

Additionally, international cooperation uncertainties arise from Southeast Asia's geopolitical centrality. U.S.-China rivalries, via strategies like the Indo-Pacific Framework and BRI, entangle climate efforts [42]. Climate issues, intersecting politics and economics, risk co-optation; colonial legacies heighten wariness, amplifying volatility [43].

10.4.2 National Level Challenges

Nationally, challenges include developmental disparities and sub-regional cooperation gaps.

ASEAN's economic heterogeneity—Singapore and Brunei's per-capita GDP exceeds $30,000, while Laos and Cambodia lag—hampers uniform governance [44]. Resource allocation varies: advanced economies invest in tech, but others prioritize growth over mitigation [43]. Compared to emitters like China and India, ASEAN's techno-economic capacity limits ambition [45].

Sub-regional cooperation deficiencies exacerbate this. Indonesia's Omnibus Law prioritizes jobs over environment, reflecting growth biases that overlook long-term sustainability [46]. Transboundary risks, like haze from peat fires, demand coordination, yet obligations remain undefined, fostering inconsistencies [47]. This fragmentation undermines collective resilience, as climate impacts cascade regionally.

10.5 Strategies for Response

Addressing these challenges requires targeted strategies, blending continuity with reform at international and national levels.

Reforming the ASEAN Way

ASEAN should evolve the "ASEAN Way" flexibly, introducing supervisory mechanisms and binding agreements [48]. A unified legal framework could set phased targets, with regional task forces for assessments [49]. Incentive-punishment systems would enhance accountability, drawing from EU models while preserving consensus [50].

Navigating Geopolitical Influences

Amid rivalries, ASEAN must adopt hedging strategies to avoid marginalization [51]. Balancing U.S.-Japan influences with China's rise, it exhibits "dual-pricing," potentially tilting westward [52]. Climate cooperation will advance circuitously, requiring diplomatic agility to secure benefits without dependency [53].

Formulating Differentiated Governance Targets

Given disparities, uniform standards are infeasible. Inspired by the Paris Agreement's bottom-up NDCs, ASEAN should tailor targets, using EU-like per-capita GDP calculations with adjustments for equity [54]. Clear legal obligations and mechanisms would bolster enforcement [55].

Strengthening Intra- and Extra-Regional Cooperation

Funding and technology bottlenecks persist; under CBDR, developed nations must provide aid [56]. ASEAN should leverage UNFCCC mechanisms for capacity building [57]. Intra-regionally, enhance sub-regional ties via information sharing, reducing costs through comparative advantages [58]. Extra-regionally, partnerships with the EU, UN, and ADB can amplify efforts, as seen in the ASEAN-EU Plan of Action [57].

10.6 Multi-Level Governance Perspective on ASEAN Climate Governance

Traditional governance models, often state-centric and hierarchical, have proven inadequate for addressing the multifaceted, transboundary nature of climate change in ASEAN. A multi-level governance (MLG) perspective offers a more inclusive framework, emphasizing coordination across international, regional, national, subnational, and non-state actors [59]. This approach recognizes climate change as a "wicked problem" requiring adaptive, polycentric strategies that leverage diverse capacities [60]. In ASEAN, MLG can bridge sovereignty concerns with collaborative imperatives, enhancing efficacy amid escalating risks. This section analyzes climate impacts on ASEAN, the rationale for MLG, ASEAN's positioning, and optimization pathways.

10.6.1 Adverse Impacts of Climate Change on ASEAN Countries

ASEAN's tropical and subtropical geography amplifies climate vulnerabilities, manifesting in widespread ecological, economic, and social disruptions. Sea-level rise, driven by glacial melt and thermal expansion, threatens coastal populations: nations like Indonesia (an archipelago of over 17,000 islands) and the Philippines face inundation of low-lying areas, with projections estimating 36 Indonesian islands could vanish by 2060 [61]. Saltwater intrusion contaminates aquifers, affecting agriculture in Vietnam's Mekong Delta, a global rice exporter [62].

Extreme weather intensification—typhoons, floods, and droughts—exacerbates these threats. The region endures annual typhoons like Haiyan (2013), which killed over 6,000 in the Philippines, with climate change increasing storm intensity by 10–20% [63]. Floods and droughts disrupt food systems: crop failures could reduce rice yields by 50% in some areas, imperiling food security for 200 million [64]. Biodiversity loss is acute; Southeast Asia hosts 20% of global species, yet warming accelerates habitat degradation. Indonesia's rainforests, a carbon sink absorbing 2 billion tons annually, suffer from wildfires linked to El Niño, releasing emissions equivalent to the EU's yearly total [65]. Coral bleaching in the Coral Triangle endangers fisheries supporting 120 million livelihoods [66].

Socioeconomically, impacts deepen inequalities: rural poor in Laos and Cambodia, reliant on subsistence farming, face disproportionate losses, potentially displacing 10–15 million by 2050 [67]. Urban centers like Jakarta and Bangkok grapple with subsidence and flooding, straining infrastructure [68]. Health risks rise with vector-borne diseases like dengue, projected to affect 50 million more annually [57]. These cascading effects underscore ASEAN's urgency for MLG, as unilateral actions falter against interconnected risks.

10.6.2 Importance of Multi-Level Governance

MLG is vital for climate response due to the issue's complexity, transcending national boundaries and requiring integrated efforts. Unlike top-down models, MLG fosters vertical (e.g., global-regional-national) and horizontal (e.g., cross-sectoral) coordination, mobilizing diverse actors for resilient outcomes. Globally, MLG facilitates consensus amid divergent interests: the Paris Agreement's NDCs exemplify bottom-up contributions, complemented by top-down oversight. Nationally, it empowers subnational entities; local governments implement context-specific adaptations, like Manila's flood defenses. Stakeholder inclusion—governments, businesses, NGOs, communities—enhances innovation: enterprises drive green tech, NGOs monitor compliance, and citizens adopt sustainable behaviors.

In ASEAN, MLG yields political (e.g., enhanced regional voice), economic (e.g., low-carbon growth), and environmental (e.g., transboundary management) benefits. It aligns with the "ASEAN Way" by enabling flexible networks, addressing gaps in centralized approaches. For instance, subregional initiatives like the Greater Mekong Subregion program integrate climate into development, demonstrating MLG's potential for synergy.

10.6.3 ASEAN's Role and Positioning in Climate Governance

ASEAN occupies a dual role as climate victim and governance participant, shaping its hybrid positioning in global and regional arenas. As a developing bloc, it advocates CBDR, emphasizing historical responsibilities of industrialized nations while safeguarding growth rights. Internationally, ASEAN engages via UNFCCC, ratifying key accords and issuing joint statements. The 2007 ASEAN Declaration on Environmental Sustainability and 2009 Joint Statement on Climate Change articulate positions, bridging North–South divides. Hosting COP13 in Bali elevated its profile, influencing the Bali Roadmap. ASEAN's discourse stresses equity, per capita emissions, and development space, countering uniform obligations. Regionally, ASEAN coordinates via AWGCC and integrates climate into ASCC blueprints. Instruments like APAEC, the Transboundary Haze Agreement, and Peatland Strategy form a normative network. Norms of sovereignty and mutual benefit guide "soft institutionalism," fostering gradual consensus amid diversity. Yet, consensus limits binding force, contrasting EU supranationalism.

ASEAN's centrality in Asia–Pacific forums amplifies influence, but geopolitical tensions challenge autonomy. Overall, it positions as a pragmatic mediator, balancing vulnerability with agency.

10.6.4 Optimization Strategies Under Multi-Level Governance

To leverage MLG, ASEAN must optimize institutions, resources, participation, and cooperation: First, enhance coherence: integrate climate across sectors via ministerial mechanisms, mainstreaming into economic plans; Second, mobilize resources: access GCF and GEF funds, expand ACGF for green finance, and foster PPPs for renewables; Third, empower subnational: support CBA in reforestation and agriculture, building local resilience; Fourth, expand partnerships: deepen EU ties for tech transfer, UN collaborations for capacity, and South-South exchanges; Finally, establish M&E: standardized data and periodic reviews, evolving toward "soft enforcement" with incentives. These strategies can transform ASEAN's governance, fostering adaptive, inclusive responses.

10.7 Case Studies: National Approaches Within ASEAN

National variations within ASEAN illustrate the practical application of multi-level governance (MLG) principles, highlighting the tensions between economic growth imperatives, climate commitments, and adaptive strategies. As developing economies, ASEAN members like Indonesia and Thailand face unique challenges in balancing rapid industrialization with vulnerability to climate impacts. These case studies delve into their policy frameworks, institutional mechanisms, implementation hurdles, and forward-looking pathways, drawing on recent developments up to 2025. By examining these nations, we can discern how MLG—encompassing national policies, subnational initiatives, private sector involvement, and international partnerships—can bridge gaps in capacity and ambition. Indonesia, as the region's largest emitter and economy, exemplifies the struggle with deforestation and coal dependency, while Thailand's energy-intensive growth underscores transition toward renewables amid policy instability.

10.7.1 Indonesia: Balancing Economic Growth and Climate Commitments

Indonesia, the world's largest archipelagic state and ASEAN's most populous nation with over 275 million people, stands at the epicenter of global climate discourse due to its dual role as a major greenhouse gas emitter and a highly vulnerable country. Its emissions profile is dominated by land-use changes, particularly deforestation and peatland fires, which account for approximately 60% of its total emissions, far exceeding energy sector contributions in most nations. The country's vast rainforests, often dubbed the "lungs of the Earth," serve as critical carbon sinks, but rampant conversion for palm oil, mining, and agriculture has led to annual emissions equivalent to those of the entire European Union. In 2025, Indonesia's climate governance reflects a complex interplay of ambitious pledges, policy reforms, and persistent implementation challenges, all within the framework of its Nationally Determined Contribution (NDC) under the Paris Agreement.

Indonesia's climate policy evolution traces back to its ratification of the UNFCCC in 1994 and the Kyoto Protocol in 2004, but significant momentum built post-2007 with the hosting of COP13 in Bali, which positioned the country as a global advocate for forest-based mitigation. The 2009 National Action Plan on Greenhouse Gas Emissions Reduction (RAN-GRK) set initial targets, followed by the 2016 ratification of the Paris Agreement. By 2025, Indonesia's updated NDC commits to an unconditional 29% reduction in emissions below business-as-usual levels by 2030, escalating to 41% with international support, and aims for net-zero emissions by 2060. This aligns with the Long-Term Strategy for Low Carbon and Climate Resilience (LTS-LCCR 2050), which emphasizes forestry as a pathway to carbon neutrality, projecting forests to absorb 140 million tons of CO2 equivalent annually by 2060. Key initiatives include the Forest and Other Land Use (FOLU) Net Sink 2030 plan, introduced at COP26 in 2021, which seeks to transform forests from emission sources to sinks through reforestation and peatland restoration.

Institutionally, Indonesia employs a multi-level approach, with the Ministry of Environment and Forestry (KLHK) leading coordination via the National Council on Climate Change. Subnationally, provincial governments play pivotal roles; for instance, Central Kalimantan's peatland restoration projects engage local communities in fire prevention and sustainable livelihoods, exemplifying MLG's bottom-up elements. Private sector involvement is bolstered through mechanisms like the Indonesia Sustainable Palm Oil (ISPO) certification, aimed at curbing deforestation in the $40 billion palm oil industry. International partnerships amplify these efforts: the $20 billion Just Energy Transition Partnership (JETP) with G7 nations, launched in 2022, funds coal phase-outs and renewable expansions, while Norway's $1 billion REDD + (Reducing Emissions from Deforestation and Forest Degradation) pledge supports monitoring and enforcement.

Despite these advancements, Indonesia faces formidable challenges. Coal remains entrenched, powering 60% of electricity with plans for 13 gigawatts of new capacity by 2030, contradicting net-zero goals and exacerbating air pollution in urban centers like Jakarta. The 2020 Omnibus Job Creation Law, while streamlining investments, relaxed environmental impact assessments, leading to increased deforestation rates and international backlash, including from the EU's impending carbon border adjustment mechanism (CBAM), which could impose tariffs on high-emission Indonesian exports. Vulnerability is acute: sea-level rise threatens 42 million coastal residents, while extreme weather, such as the 2024 floods in Sumatra, displaces communities and disrupts agriculture. Geopolitically, climate security intersects with territorial disputes in the South China Sea, where rising seas could erode maritime boundaries, challenging sovereignty. Moreover, investment gaps persist: achieving renewable transformation requires $65 billion from 2025–2030, scaling to $1.08–1.22 trillion by 2050, yet domestic financing covers only 20%. Policy inconsistencies, such as subsidies for fossil fuels totaling $20 billion annually, undermine transitions.

In an MLG context, opportunities abound for optimization. Enhancing subnational capacities through decentralized funds could empower provinces like Papua in biodiversity conservation, linking to global carbon markets via voluntary mechanisms under Article 6 of the Paris Agreement. Community-based initiatives, such as indigenous-led mangrove restoration—Indonesia's "secret weapon" with potential to sequester 3 gigatons of carbon—foster resilience and equity. Policy recommendations include streamlining financial regulations to attract green investments, revising the NDC for Paris alignment, and integrating climate into Belt and Road Initiative projects for sustainable infrastructure. Indonesia's trajectory, ranked 42nd in the 2025 Climate Change Performance Index (CCPI) as a low performer, hinges on bridging these gaps to realize its potential as a climate leader in the Global South.

10.7.2 Thailand: Energy Sector Challenges and Pathways to Improvement

Thailand, Southeast Asia's second-largest economy with a GDP of approximately $500 billion and a population of 70 million, exemplifies the challenges of transitioning an energy-intensive, export-driven economy amid escalating climate risks. Ranked 30th in the 2025 Climate Risk Index (improved from prior years), Thailand faces frequent floods, droughts, and sea-level rise, which have caused economic losses exceeding $10 billion annually in recent decades. Its climate governance, shaped by a decade of policy evolution, emphasizes carbon neutrality by 2050 and net-zero emissions by 2065, yet grapples with fossil fuel dependency and institutional fragmentation.

Thailand's policy framework has matured since ratifying the UNFCCC in 1994 and the Paris Agreement in 2016. The National Climate Change Master Plan (2015–2050) laid early foundations, evolving into the 2021–2030 Climate Change Master Plan, which integrates mitigation across sectors like energy, agriculture, and transport. By 2025, the impending Climate Change Act—nearing parliamentary approval—establishes a legal framework for low-carbon transitions, mandating carbon pricing, emissions reporting, and incentives for renewables. The Power Development Plan (PDP) 2024–2037 targets 50% renewable energy by 2037, up from 20% in 2024, with emphases on solar, wind, and bioenergy. The National Adaptation Plan (NAP) addresses vulnerabilities in six sectors, including water management and agriculture, supported by the Thailand Country Programme on Climate Change, which outlines strategies for resilience building.

MLG is evident in Thailand's structure: the Office of Natural Resources and Environmental Policy and Planning (ONEP) coordinates nationally, while provincial administrations implement localized adaptations, such as Bangkok's flood barriers under the Bangkok Master Plan on Climate Change. Private sector engagement thrives through initiatives like the Thailand Carbon Neutral Network, involving corporations in voluntary reductions. International collaborations, such as the UNDP's SCALA program, focus on climate-smart agriculture, attracting private investments for resilient farming in the Northeast region. The ASEAN Power Grid interconnection enhances regional energy security, allowing imports of hydropower from Laos.

Challenges abound, however. Fossil fuels comprise 75% of energy, with natural gas at 60%, subsidized at $5 billion yearly, distorting markets and delaying renewables. Policy delays—the Climate Change Act, drafted in 2019, remains pending amid political transitions—create uncertainty for investors. Vulnerability is pronounced: the 2024 floods in Chiang Mai caused $1 billion in damages, while droughts in the Chao Phraya basin threaten rice exports, valued at $4 billion annually. Human capital shortages in green tech and overlapping agencies (e.g., Energy Ministry and ONEP) hinder coordination. Geopolitically, reliance on imported fuels exposes Thailand to global price volatility, exacerbated by climate-induced disruptions in supply chains.

Pathways forward leverage MLG: redirect subsidies to renewables, potentially unlocking $100 billion in investments by 2030 for high-income status. Subnational pilots, like Chonburi's EV hubs, can scale nationally. Enhancing the PDP with enforceable targets and fostering East Asian tech transfers could accelerate transitions. Thailand's 2025 CCPI ranking as a medium performer signals progress, but realizing net-zero demands integrated, multi-level action.

10.7.3 ASEAN's Cooperation with International Organizations

ASEAN's collaboration with international organizations in climate governance has been instrumental in shaping its regional strategies, providing technical expertise, financial resources, and normative guidance. Since the early 1990s, these partnerships have evolved from ad hoc, project-based engagements to more structured, multi-dimensional frameworks, reflecting ASEAN's transition from a passive recipient of aid to an active participant in global climate diplomacy. This evolution aligns with key milestones in international climate regimes, such as the UNFCCC (1992), Kyoto Protocol (1997), and Paris Agreement (2015), where ASEAN has sought to leverage external support to address its vulnerabilities while asserting regional centrality. Rooted in principles like common but differentiated responsibilities (CBDR), these collaborations encompass capacity building, policy harmonization, and joint projects, yet they also pose risks of dependency and unequal power dynamics. This section examines ASEAN's ties with major partners—the United Nations system, the European Union, and the Asian Development Bank—highlighting achievements, mechanisms, and challenges, while drawing on multi-level governance perspectives to underscore synergies across scales.

Early engagements with the United Nations (UN) system laid the foundational architecture for ASEAN's climate cooperation. Following the 1992 Rio Earth Summit, ASEAN members ratified the UNFCCC, marking their entry into the global climate regime. The UN Environment Programme (UNEP) and United Nations Development Programme (UNDP) played pivotal roles in initial capacity building. For instance, the ASEAN-UNDP Climate Change Adaptation Initiative (CCAI), launched in the late 1990s, focused on vulnerability assessments and policy consultations, enhancing regional data systems and adaptive planning in flood-prone areas like the Mekong Delta. Similarly, UNEP's Regional Resource Center for Asia and the Pacific (RRC.AP), established in Bangkok in 1999, provided technical assistance for emissions inventories and environmental monitoring, supporting ASEAN's early responses to transboundary issues such as haze pollution. These initiatives were largely externally driven, addressing ASEAN's limited internal resources during the pre-2007 phase, but they fostered information exchange among member states, aligning with MLG's emphasis on vertical coordination between global and regional levels.

Post-Kyoto developments deepened these ties, injecting momentum into ASEAN's institutional frameworks. The Kyoto Protocol's entry into force in 2005 prompted ASEAN to formalize partnerships, culminating in the 2007 ASEAN-UN Environmental Cooperation Framework. This agreement emphasized joint responses to climate change and natural disasters, integrating UN expertise into ASEAN's Socio-Cultural Community pillar. Building on this, the 2010 ASEAN-UN Action Plan on Environment and Climate Change expanded scopes to include renewable energy promotion, environmental education, and disaster risk reduction, with specific projects like the UN's support for the ASEAN Agreement on Disaster Management and Emergency Response (AADMER). By 2015, as the Paris Agreement loomed, ASEAN-UN collaboration matured into a multi-layered network: ministerial dialogues facilitated high-level policy alignment, while technical cooperation—such as UNDP's assistance in Nationally Determined Contributions (NDCs)—bolstered national implementation. The UN Economic and Social Commission for Asia and the Pacific (UNESCAP) further enhanced this by promoting regional data-sharing and sustainable development goals (SDGs) integration, helping ASEAN countries like Vietnam and the Philippines develop climate-resilient infrastructure. These efforts have yielded tangible outcomes, such as improved early warning systems that mitigated losses during the 2024 typhoon season, demonstrating MLG's efficacy in linking global norms with local actions.

Parallel to UN engagements, ASEAN's partnership with the European Union (EU) represents one of its most comprehensive and influential collaborations, focusing on norm diffusion, institutional learning, and green transitions. Originating in the early 1990s with environmental protection dialogues, the partnership gained strategic depth with the 2007 ASEAN-EU Dialogue on Sustainable Development, which elevated climate issues to a core agenda. The 2018–2022 ASEAN-EU Plan of Action further institutionalized this, emphasizing practical cooperation in emissions trading, renewable energy, and circular economies. Central to this is the Enhanced Regional EU-ASEAN Dialogue Instrument (E-READI), launched in 2015, which facilitates thematic dialogues on climate policy, such as workshops on carbon pricing and energy efficiency. The EU's advanced frameworks, like the Fit for 55 packages, have inspired ASEAN's ambitions: for example, Singapore's carbon tax draws from the EU Emissions Trading System (ETS), while joint projects under the EU's Global Gateway initiative fund low-carbon infrastructure in Indonesia and Malaysia. Financially, the EU has committed over €2 billion since 2010 for ASEAN climate programs, including green finance mechanisms that support small and medium enterprises (SMEs) in adopting sustainable practices. This collaboration extends to MLG by involving subnational actors: EU-funded city-to-city partnerships, such as between Amsterdam and Jakarta, promote urban resilience through flood management and smart grid technologies. However, divergences arise from the EU's stringent standards, such as the Carbon Border Adjustment Mechanism (CBAM), which could impose tariffs on ASEAN exports, prompting calls for technology transfers to ensure equitable transitions.

The Asian Development Bank (ADB) complements these efforts by focusing on infrastructure resilience and financial mobilization, addressing ASEAN's adaptation needs in a region where climate damages could reach 11% of GDP by 2100. Since the 1990s, ADB has integrated climate into its lending, with the 2009 Economics of Climate Change in Southeast Asia report highlighting vulnerabilities and advocating for $40 billion in annual investments. By 2025, ADB has disbursed over $10 billion for ASEAN adaptations, funding projects like the Greater Mekong Subregion's climate-resilient transport corridors and Vietnam's coastal protection schemes. The ASEAN Catalytic Green Finance Facility (ACGF), co-launched with ADB in 2019, mobilizes $2 billion for green infrastructure, blending public and private funds to de-risk investments in renewables. ADB's support extends to policy advisory, such as assisting Thailand's Power Development Plan revisions for net-zero alignment. In MLG terms, ADB emphasizes community-level interventions: programs like the Community Resilience Partnership Program Trust Fund engage local governments in Cambodia and Laos for grassroots adaptation, fostering bottom-up innovation.

These collaborations enhance ASEAN's resources but introduce challenges, including dependency risks and normative mismatches. External funding, while crucial, can skew priorities toward donor agendas, as seen in EU-driven carbon markets that may overlook ASEAN's developmental needs. Geopolitical tensions, such as U.S.-China rivalries, complicate partnerships, with ASEAN wary of being instrumentalized in broader power plays. To mitigate, ASEAN must assert centrality, as articulated in the 2021 ASEAN Outlook on the Indo-Pacific, by negotiating equitable terms and building internal capacities. Future prospects lie in expanding ties with entities like the Green Climate Fund (GCF), where ASEAN has accessed $500 million for mitigation since 2015, and fostering South-South cooperation with partners like China under the BRI's green pillar. Ultimately, these partnerships, when aligned with MLG, can propel ASEAN toward resilient, sustainable governance, bridging global ambitions with regional realities.

10.8 Conclusion

Over three decades, ASEAN's climate governance has progressed from exploratory inclusion to institutionalized frameworks, yet challenges like the "ASEAN Way's" limitations and geopolitical uncertainties persist. MLG offers a pathway forward, integrating diverse actors for resilient, equitable responses. Case studies underscore national adaptations, while international ties amplify capacities. As global temperatures rise, ASEAN must reform internally and engage externally to secure sustainable futures, contributing to broader climate order.

References

1. Asian Development Bank (2023) Delaying climate action will cost Southeast Asia dearly. http://www.adb.org/news/southeast-asia-faces-soaring-economic-costs-if-climate-change-action-delayed-new-study-zh. Accessed 11 Jul 2024
2. International Monetary Fund (2007) The economics of climate change. http://www.imf.org/external/pubs/ft/survey/so/2007/POL1026A.htm. Accessed 23 Oct 2024
3. Chenghui X, Li, Zhijiang (2012) ASEAN countries' climate governance cooperation: status and issues. Yuejiang J 6:83–89. https://doi.org/10.13878/j.cnki.yjxk.2012.06.002
4. Zhang P (2015) Analysis of ASEAN's climate change policies. Master's thesis, Huazhong University of Science and Technology
5. Ji L (2016) Security concepts and ASEAN's climate change cognition and policies. Int Secur Res 3:114–130, 154–155. https://doi.org/10.14093/j.cnki.cn10-1132/d.2016.03.006

6. Council of the European Union (2023) Fit for 55. https://www.consilium.europa.eu/en/policies/green-deal/fit-for-55-the-eu-plan-for-a-green-transition/
7. Zhao Q (2021) Strengthening ASEAN's coordination capacity in addressing climate change. China Social Sciences Daily. 6 Dec 2021
8. World Bank (2023) GDP data. https://data.worldbank.org.cn/indicator/NY.GDP.MKTP.CD?name_desc=false. Accessed 23 Dec 2024
9. ASEAN Environment Cooperation (2023) Statements & declarations. http://environment.asean.org/statements-and-declarations/. Accessed 24 Nov 2024
10. Ying C, Qingchen C (2022) Evolution trends and countermeasures for global climate risks. People's Forum Acad Ed 14:20–23
11. Xiaoguang X, Du, Xiaojie (2022) Arctic shipping safety governance: multi-level governance, dilemmas, and pathways. J Ocean Univ China (Soc Sci) 3:68–77
12. Meichao J, Xu, Yueting (2020) Analysis of internalization differences in EU climate governance rules from a multi-level governance perspective. Soc Sci Forum 5:197–210
13. Wu, Jing, Qianting Z (2020) Analysis of cities' role in global climate governance in the post-Paris Agreement era. Environ Prot 48(5):12–17
14. Kai S (2017) Mechanism changes, multi-level governance, and the future of Arctic governance. Foreign Aff Rev 3:109–129
15. Guo Z (2022) Global perspective: Indonesia's renewable energy reserves and development. China Power Media Data R&D Center
16. BP (2023) Statistical review of world energy. https://www.bp.com/en/global/corporate/energy-economics/statistical-review-of-world-energy.html. Accessed 29 Jun 2024
17. Hasan MH, Mahlia TMI, Nur H (2012) A review on energy scenario and sustainable energy in Indonesia. Renew Sustain Energy Rev 16(4):2317–2326
18. Vidal J (2014) Rate of deforestation in Indonesia overtakes Brazil, says study. The Guardian, June 30, 2014.
19. Gu, Yongqiang (2010) Indonesia's new energy development momentum is strong. China Petrochem 2:45–48
20. Xinhua T, Fanggui Q (2015) Analysis of Indonesia's mining investment environment and related legal systems. Southeast Asia Stud 3:55–56
21. Jiaxin L (2014) Sorting and trend analysis of Indonesia's export policies. China Powder Ind 1:1–2
22. Juan'er Z, Guohua Y, Shihu W (2014) Study on the impact of changes in Indonesia's mining laws and policies on China. China Land Resour Econ 5:46–50
23. Mujiyanto S, Tiess G (2013) Secure energy supply in 2025: Indonesia's need for an energy policy strategy. Energy Policy 61:31–41
24. Economic and Commercial Counselor's Office, Embassy of the PRC in Indonesia (2014) Indonesia's renewable energy plan requires $36 billion in funding. http://id.mofcom.gov.cn/article/ziranziyuan/zhengt/201411/20141100799687.shtml
25. China-Indonesia Energy Cooperation Report (2023) https://www.goalfore.cn/a/3562.html. Accessed 21 Mar 2024
26. Economic and Commercial Counselor's Office, Embassy of the PRC in Indonesia (2013) Indonesia to introduce new regulations to encourage solar development. http://id.mofcom.gov.cn/article/ziranziyuan/zhengt/201302/20130200037691.shtml
27. 39th ASEAN Energy Ministers' Meeting (2023) Joint declaration on security and energy transition in Bandar Seri Begawan. ASEAN Secretariat
28. ASEAN Plus Three Energy Ministers' Meeting (2009) Strengthening cooperation in energy. http://www.gov.cn/jrzg/2009-07/30/content_1379058.htm. Accessed 26 Mar 2024
29. ASEAN Secretariat (1997) ASEAN Vision 2020. https://asean.org/?static_post=asean_vision-2020. Accessed 2 Jul 2024
30. China Overseas Industry Development Association (2023) http://www.ciodpa.org.cn. Accessed 7 Jul 2024
31. IEA (2023) An energy sector roadmap to net zero emissions in Indonesia. https://www.iea.org/reports/an-energy-sector-roadmap-to-net-zero-emissions-in-indonesia/executive-summary. Accessed 20 Nov 2024

32. Eckstein D et al (2021) Global climate risk index 2021. Germanwatch. https://www.germanwatch.org/en/19777. Accessed 26 Sept 2024
33. Xinhua Net (2014) Report claims thousands of Indonesian islands could disappear in 36 years due to sea-level rise. http://news.xinhuanet.com/travel/2014-02/27/c_126197975.htm. Accessed 29 Oct 2024
34. Think-Asia (2009) The economics of climate change in Southeast Asia: a regional review. https://think-asia.org/handle/11540/179. Accessed 16 Oct 2024
35. EU Publications (2023) GHG emissions of all world countries 2023 report. https://op.europa.eu/en/publication-detail/-/publication/0cde0e23-5057-11ee-9220-01aa75ed71a1/language-en. Accessed 26 Oct 2024
36. Xu X (2023) Indonesia's participation in global climate governance: motivations and limits. Strat Decis Res 3:42–60, 106–107
37. Wei H, Lai C (2022) Indonesia's Jokowi government's climate policies and China-Indonesia climate cooperation. Reg Glob Dev 5:95–118, 158
38. Xiuju S, Pengcheng W (2018) Indonesia's environmental disasters and participation in global environmental governance. Southeast Asia Stud 3:3–9
39. Customs Information Network (2023) China Customs Administration. Data Source
40. World Bank (2023) http://data.worldbank.org.cn. Accessed 24 Nov 2024
41. Nansheng Jinshi Shuo (2022) Thailand's 2021 GDP ranks second in ASEAN, what about the Philippines, Malaysia, Vietnam? Sina Finance
42. World Bank (2023) EG.GDP.PUSE.KO.PP.KD. http://data.worldbank.org.cn/indicator/EG.GDP.PUSE.KO.PP.KD. Accessed 26 Nov 2024
43. Huang D (n.d.) Analysis of climate change policy, pp 197–199
44. Huang D, Ji L (2020) The climate community: the ASEAN way to address climate change in the post-Paris era
45. ASEAN Energy Center (2020) ASEAN climate action plan: a brief review of independent contributions of ASEAN countries
46. Xi W, Yuan Y (2021) Analysis and cooperation suggestions on policy mechanisms on climate change in ASEAN countries
47. Ji L (2016) Security concept and ASEAN climate change cognition and policy
48. Thailand Reduces Dependence on Natural Gas (2022) Thailand reduces dependence on natural gas for power generation, low-carbon electricity promotes renewable energy development. China Southern Power Grid Newspaper. 2022
49. Distributed Energy Network (2022) Low-carbon electricity in Thailand promotes renewable energy development
50. Zhang R, Wang X (2022) ASEAN energy transformation: dilemmas and outlook
51. Ministry of Energy, Thailand (2023) Thailand Energy Ministry official website
52. Zuo X, Xi F, Liu J (2022) Thailand's new energy policy and implications for Chinese enterprise investment
53. Southern Energy Observation (2023) Many ASEAN countries set clear wind power targets, power grid construction needs acceleration
54. Yuan Z, Jiang S (2021) ASEAN energy governance: a perspective of mechanism change
55. Zhou X (2023) East Asian wisdom contributes to global green development
56. Dessler J, Parson EA (2012) Climate change: science or politics? China Environment Press, Beijing, pp 54–68
57. Wang R (2024) Western international climate cooperation concepts and global climate governance dilemmas. Int Stud 6:89–106, 130–131
58. Yu B (2023) Power discourse practice: logic of deep changes in global governance. Shanghai People's Publishing House, Shanghai
59. Ma, Mingfei, Xinrui W (2024) Examination and reflections on losses and damages caused by climate change. Environ Prot 52(Z3):74–78
60. Yitian H (2021) Dual impacts of international manufacturing division on developing countries' emission reduction policies: an analytical framework. Fudan Int RelatS Rev 2:115–134

61. Han R (2023) Comprehensive assessment methods for climate governance policies and their applications. Tsinghua University Press, Beijing
62. Fan J (2010) Geopolitical games in global climate governance. Eur Stud 28(6):1–18, 158
63. Fei W, Chen C, Pengfei L (2024) Evolutionary logic, causal analysis, and pathways for global climate security issues: an empirical study with the UN. Environ Prot 52(16):68–73
64. Weiguo W, Dehai L, Wenyan S et al (2014) Game tree correlated equilibrium analysis of Copenhagen international climate negotiations. Syst Eng Theory Pract 34(7):1697–1704
65. Yuxuan W (2025) Global climate governance: Trends, challenges, and pathways. Soc Sci Front 4:277–280
66. Xie M (2025) Network effects of international environmental agreements and domestic climate governance actions. World Econ Politics 4:133–163, 168
67. Yuemin H (2024) Ecological interest games in global climate governance and China's countermeasures. Qinghai Soc Sci 1:38–48
68. Bingyu L (2024) Investment arrangements for climate governance: origins, evolution, and new pathways. Law Bus Res 41(4):55–70

Chapter 11
Coalitions of Change: State Alliances and Shaping of Global Climate Politics

This chapter explores the intricate formation, evolution, and interactions of state coalitions within the framework of global climate governance, from the foundational Rio Earth Summit in 1992 to the transformative outcomes of recent COP summits up to 2025. By examining key groups such as the G77 + China, the Umbrella Group, OPEC, AOSIS, and LDCs, it elucidates how these alliances navigate the tensions between equity, vulnerability, economic interests, and institutional constraints under the UNFCCC. Drawing on historical analyses of the Conference of the Parties (COPs), the discussion highlights the shift from rigid North–South dichotomies to multipolar, fragmented dynamics, where coalitions amplify voices, contest norms like CBDR, and drive both progress and impasses in addressing the climate crisis. Ultimately, this overview underscores the role of group politics in reshaping multilateralism, offering insights into future pathways for inclusive, adaptive governance in an era of poly-crisis.

11.1 Group Politics and the Institutional Tensions in Climate Governance

The landscape of global climate governance, spanning from the Rio Earth Summit in 1992 to the Glasgow Climate Pact of 2021, has been profoundly shaped by the phenomenon of group politics. This chapter delves into the intricate web of state coalitions that have emerged as pivotal actors in international climate negotiations. Rather than viewing states as isolated entities pursuing narrow national interests, we must recognize how they aggregate into coalitions—formal and informal alliances that amplify collective bargaining power, articulate shared identities, and navigate the complex interplay of economic, ecological, and ethical imperatives [1]. These coalitions, ranging from the expansive "Group of 77 plus China" (G77 + China) to the defensively oriented Umbrella Group, the energy-centric Organization of

X. Zou, *Thirty Years of Global Climate Governance*, Energy, Climate Change & Sustainability in the Global South, https://doi.org/10.1007/978-981-95-8457-4_11

the Petroleum Exporting Countries (OPEC), and the vulnerability-driven Alliance of Small Island States (AOSIS) and Least Developed Countries (LDCs), embody the structural tensions inherent in a multilateral system grappling with asymmetric responsibilities and divergent vulnerabilities.

At its core, group politics in climate governance arises from the inherent inequalities of the international system [2]. Climate change is not merely an environmental challenge; it intersects with developmental rights, survival imperatives, energy dependencies, and economic redistributions, thereby refracting the longstanding North–South divide into new configurations. The formation of these coalitions is a strategic response to the institutional architecture of the United Nations Framework Convention on Climate Change (UNFCCC), where consensus-based decision-making often favors numerical majorities or moral suasion over raw power. As such, coalitions serve as mechanisms for agenda-setting, responsibility allocation, and legitimacy production, transforming abstract principles like "common but differentiated responsibilities" (CBDR) into contested battlegrounds [3].

This chapter draws extensively from a rich array of historical documents, scholarly analyses, and empirical case studies embedded in the provided materials. It begins with a detailed examination of the G77 + China, tracing its trajectory from unified Southern solidarity to internal fragmentation amid evolving global dynamics. Subsequent sections analyze the Umbrella Group and OPEC as exemplars of defensive strategies employed by developed and resource-dependent states, respectively. We then explore the "politics of vulnerability" through the lenses of AOSIS and LDCs, illustrating how marginalized actors leverage moral authority to migrate from the periphery to the center of global agendas. Finally, the chapter synthesizes the interactive patterns among these coalitions and reflects on their institutional futures, offering theoretical insights into the dialectics of confrontation and cooperation in climate multilateralism. By unpacking these dynamics, this analysis reveals how group politics has not only driven the evolution of climate regimes but also exposed the limitations of multilateralism in addressing planetary crises. From the pinnacle of Rio's optimism to the fragmented compromises of Glasgow, coalitions have been the engines of progress and impasse, underscoring the need for a more inclusive and adaptive governance framework in the Anthropocene era.

The materials underscore the G77 + China's foundational role as a mechanism for enhancing Southern voices in UN economic deliberations, established in 1964 with an initial 77 members and expanding to over 120 by 1979 while retaining its name [4]. Its loose organizational structure—lacking a headquarters, charter, or budget—relies on consensus and ministerial meetings to coordinate positions, reflecting a flexible yet fragile unity. In climate contexts, this coalition has navigated from the CBDR's enshrining in 1992 to the Paris Agreement's reinterpretation in 2015, where universal mitigation obligations blurred traditional divides. Similarly, the Umbrella Group's informal nature contrasts with the EU's institutionalized approach, while OPEC's energy politics introduces paradoxes of dependency and resistance. AOSIS and LDCs, meanwhile, highlight existential stakes, with the former's 1990 formation driven by sea-level threats and the latter's 46 members bearing disproportionate impacts despite minimal emissions. This chapter sets the stage for a nuanced exploration, emphasizing how these coalitions' evolutions mirror broader shifts in global power, from bipolar Cold War remnants to multipolar fragmentation, and how they continue to shape the quest for climate justice.

11.2 The G77 + China: From South-South Solidarity to Internal Differentiation

The G77 + China represents the most populous and heterogeneous coalition in global climate negotiations, embodying the aspirations and contradictions of the Global South. Comprising over 130 developing countries, including China as a key partner, this alliance has evolved from a beacon of unified resistance against Northern dominance to a fragmented entity grappling with internal divergences [5]. Its history, as detailed in the materials, illustrates the challenges of maintaining cohesion amid diverse economic structures, ecological vulnerabilities, and developmental trajectories.

11.2.1 Origins and Peak of Cooperation (1992–1997)

The roots of the G77 + China in climate governance trace back to the broader Group of 77, formed in 1964 during the first United Nations Conference on Trade and Development (UNCTAD). Initially comprising 77 developing nations, it aimed to articulate collective economic interests, enhance bargaining power in North–South dialogues, and foster South-South cooperation. By the 1990s, as climate change emerged as a critical global issue, China aligned with this group to form the G77 + China mechanism, a strategic partnership that amplified Southern influence in UNFCCC negotiations [6].

The 1992 Rio Earth Summit marked the alliance's zenith. Here, developing countries, united under the G77 + China banner, successfully advocated for the CBDR principle in the UNFCCC. This principle acknowledged historical emissions from industrialized nations while exempting developing countries from binding reduction targets, framing climate action as an equity issue rather than a universal burden. The materials highlight how this "positive discrimination" reflected tight collaboration, with China and other emerging economies emphasizing development rights over immediate mitigation. The alliance's diversity—encompassing small island states, least developed countries, oil exporters, and rainforest nations—paradoxically strengthened its position, allowing it to represent the broadest spectrum of Southern interests.

Organizationally, the G77 + China operates without formal institutions, relying on consensus and rotating chairmanships in New York and Geneva. This looseness facilitated early unity, as seen in the 1997 Kyoto Protocol negotiations, where the coalition collectively resisted commitments for non-Annex I countries. As noted in scholarly analyses, this period exemplified "diversity seeking unity," with the group leveraging its numerical majority to counter Northern pressures. However, even at this peak, seeds of division were sown. Subgroups like AOSIS and LDCs began voicing unique vulnerabilities, foreshadowing future splits [7]. The materials provide figures illustrating economic disparities: emerging economies like the BASIC quartet (Brazil, South Africa, India, China) accounted for significant GDP shares, while LDCs and small islands contributed minimally, highlighting uneven capacities for mitigation and adaptation [8].

11.2.2 Initial Cracks and Differentiation (2005–2015)

The mid-2000s witnessed the emergence of fissures within the G77 + China, driven by external pressures and internal heterogeneities. The 2007 Bali Action Plan's shift to "developed/developing" categorizations allowed Northern states to target "advanced developing countries" like China for enhanced obligations, eroding the alliance's monolithic front. The 2009 Copenhagen Conference crystallized these tensions. The "Danish Text" proposal, which favored a new binding framework over Kyoto's extension, divided the coalition. China and BASIC insisted on preserving Kyoto's structure, arguing for Northern leadership in reductions [9]. In contrast, AOSIS pushed for a 1.5 °C limit, viewing 2 °C as existential threats to low-lying islands. Materials detail how over 100 G77 members favored the stricter goal, clashing with emerging emitters' economic priorities [10]. The resulting Copenhagen Accord institutionalized differentiation among developing countries, marking a formal acknowledgment of South-South splits.

This era's fragmentation extended to subgroups. OPEC members, reliant on fossil exports, opposed global reductions fearing market disruptions. As per data from 2008, OPEC's GDP share was 3.51%, dwarfed by BASIC's 12.81%, yet their energy dependencies amplified resistance. LDCs, with low per capita incomes, prioritized adaptation funding, while emerging powers focused on emission spaces for industrialization. By 2015's Paris Conference, the reinterpretation of CBDR universalized mitigation, further straining unity [11]. AOSIS and LDCs joined the High Ambition Coalition with Northern powers, advocating ambitious goals and transparency. Materials from COP proceedings underscore how this cross-alliance broke the "North–South confrontation," ushering in complex multipolar dynamics [12].

11.2.3 Formal Fragmentation and Recent Developments (2015–Present)

Post-Paris, the G77 + China has trended toward fragmentation, with consensus coexisting with profound divergences across three axes: mitigation ambition, funding urgency, and finance competition [13]. Mitigation stringency varies sharply. AOSIS and LDCs, facing acute risks (e.g., sea-level rise threatening survival), demand stringent actions, as visualized in GDP-emission correlations showing their minimal contributions yet maximal vulnerabilities. Emerging powers like BASIC prioritize growth, seeking larger emission spaces; OPEC resists expansions impacting oil revenues.

Funding urgencies differ due to economic stages. LDCs and small islands, with GDPs at 1.44% globally in 2008, urgently seek aid for adaptation, supporting market mechanisms from developed states. BASIC and OPEC, with rising strengths, deprioritize funds in favor of development flexibility. Finance allocation intensifies rivalries. Kyoto's Clean Development Mechanism favored China (1,557 projects) and India (1,135), exacerbating inequities. Among BASIC, India advocates per capita allocations to constrain China; Brazil leverages renewables; South Africa flexibility for African leadership; all diverging on classification, emission standards, and long-term goals. Yet, unity resurfaces strategically. At COP27 (2022), themed "Together for Implementation," the coalition pressured for the Loss and Damage Fund, addressing unfulfilled Northern pledges [14]. Materials from the summit, including Cuban-hosted leaders' meetings, emphasize collective pushes for sustainable development, multilateralism, and justice, with China's Ding Xuexiang affirming opposition to green unilateralism. Recent events, like the 2023 Dubai summit, reinforce this. Xi Jinping's representative stressed docking green strategies, mutual aid, and defending development rights, countering technical barriers and decoupling [15].

11.2.4 Causes of Divergence: Domestic and International Factors

Employing hierarchical analysis, divergences arise domestically from ecological vulnerabilities and mitigation capacities. Vulnerability encompasses climate impacts and response effects: for AOSIS, it's survival; for LDCs, development; for OPEC, economic restructuring. Capacities depend on economic levels (enabling clean transitions) and structures (high-emission sectors like petroleum limit OPEC). Internationally, uneven incentives—finance/technology maldistribution—fuel competition. Limited aid totals (e.g., post-2020 adaptation costs at $140–300 billion annually by 2030) intensify rivalries, while varying emission pressures (higher for emerging states) polarize stances. In sum, the G77 + China's evolution from solidarity to differentiation reflects Southern diversity's double-edged nature. While fragmentation challenges cohesion, it enables nuanced advocacy, ensuring the coalition's enduring role in climate equity pursuits.

11.3 The Umbrella Group and OPEC: Institutional Defenses and Energy Politics in Developed and Resource-Dependent Coalitions

Shifting focus from Southern coalitions, the Umbrella Group and OPEC exemplify defensive maneuvers by states with high emissions and economic stakes in status quo energy systems. These groups navigate climate governance by prioritizing national interests over global imperatives, employing strategies of delay, dilution, and discourse reconfiguration.

11.3.1 The Umbrella Group: Origins, Structure, and Defensive Strategies

The Umbrella Group emerged post-Kyoto as an informal alliance of non-EU Annex I countries, including the US, Japan, Canada, Australia, New Zealand, Norway, Russia, and occasionally others [16]. Named for its "umbrella-like" geographic spread, it lacks formal structure, functioning as a venue for policy alignment to counter G77 + China demands for historical accountability. Its origins lie in Kyoto's 1997 negotiations, where members resisted bindings: the US pushed voluntary approaches; Japan and Canada emphasized developing country inclusion; Australia cited energy-export exemptions. The group's core aim—"dispersing responsibility, delaying constraints, reshaping rules"—manifests in collective resistance to CBDR's strict interpretation [17].

Early fractures appeared with US withdrawal (2001), Canadian exit (2011), and Japanese non-commitment to Kyoto's second period. Fulfillment was dismal: Canada overran by 26.7%, Australia by 20.9%, per UN data. This underscores the alliance's economic protectionism over mitigation zeal. Paris's NDC framework suited Umbrella preferences for flexibility, allowing self-defined targets. US's 26–28% cut from 2005 (14% from 1990) and Australia's similar modesty, with ongoing growth, exemplify this. Materials analyze how such goals balance growth and climate rhetoric. On finance/technology, the group diversifies sources via private capital, shares burdens with emerging economies, and upholds IP, maintaining Northern hegemony. This stance, as critiqued, perpetuates inequities by leveraging rather than funding transitions.

11.3.2 OPEC: Energy Paradox and Climate Resistance

OPEC, formed in 1960 with 13 members controlling 40% of global oil production, embodies the fossil-climate tension. Members like Saudi Arabia (99% oil–gas reliant) depend on exports for fiscal stability, viewing decarbonization as economic peril [18]. From UNFCCC's start, OPEC has doubted science, opposed bindings, and sought compensation. Materials note its 1995 Berlin influence, dominating G77 via Saudi prowess despite internal diversity (Nigeria's $450 per capita GDP vs. Qatar's $33,000). Post-Paris, OPEC reframes as "inclusive transition" advocates, with net-zero pledges (Saudi 2060) and anti-hegemony rhetoric [19]. This adapts Southern justice narratives to defend oil, emphasizing all-energy mixes. Internal rifts—geopolitical (Saudi-Iran), dependency variances—undermine unity, as 2016 reduction disputes show [20]. Scholars term it a "petroleum giant, climate dwarf," highlighting governance issues amid global importance.

11.3.3 Comparative Analysis: Northern Defense and Southern Resistance Intersections

Umbrella and OPEC defenses intersect: both resist bindings, though Umbrella via institutional flexibility, OPEC economic exemptions. Their logics sustain delayed governance, echoing Southern resistances in mutual responsibility evasion. Materials contrast OPEC's G77 influence—leveraging development legitimacy—with Umbrella's Northern core status, illustrating how defenses perpetuate North–South imbalances while fostering cross-group complicities.

11.4 The Politics of Vulnerability: From Periphery to Center in Agenda Migration

In the intricate tapestry of global climate governance, the politics of vulnerability emerges as a compelling narrative of how structurally marginalized actors—those with minimal contributions to greenhouse gas emissions yet bearing the brunt of climatic disruptions—have strategically repositioned themselves from the fringes of international negotiations to influential architects of the global agenda. This section focuses on two pivotal coalitions: the Alliance of Small Island States (AOSIS) and the Least Developed Countries (LDCs). Drawing from the provided materials, which include detailed analyses of their formation, challenges, and strategic evolutions, we explore how these groups have leveraged moral authority, ethical framing, and cross-alliance diplomacy to transform their inherent weaknesses into sources of political capital. Their journey underscores a broader shift in climate politics: from a power-centric paradigm dominated by major emitters and economic heavyweights to one increasingly attuned to equity, survival imperatives, and intergenerational justice.

Existing evidences vividly illustrate the disproportionate impacts on these vulnerable entities. For instance, small island nations and LDCs collectively represent populations that contribute negligibly to global emissions—often less than 1%—yet face existential threats from sea-level rise, extreme weather, and ecosystem degradation. This asymmetry has fueled a "victim justice" discourse, where vulnerability is not merely a liability but a moral lever to compel action from historical polluters. As the documents note, these coalitions have evolved from passive recipients of aid to proactive norm-setters, influencing key milestones like the 1.5 °C temperature goal in the Paris Agreement and the establishment of the Loss and Damage Fund at COP27. Their strategies reveal the potency of "weak state diplomacy" in multilateral forums, where numerical minorities harness narrative power to reshape institutional outcomes.

This politics of vulnerability challenges traditional international relations theories, which often prioritize material power. Instead, as constructivist perspectives suggest, these groups construct shared identities around "climate fragility," mobilizing global sympathy and pressuring dominant actors through ethical appeals. The materials provide empirical depth, detailing how AOSIS and LDCs have navigated internal diversities and external resistances to migrate agendas from peripheral concerns (e.g., adaptation funding) to central pillars of governance (e.g., global adaptation goals). In doing so, they exemplify how vulnerability can catalyze systemic change, turning the periphery into a vortex of agenda innovation.

11.4.1 AOSIS: Formation, Threats, and Development

The Alliance of Small Island States (AOSIS) stands as a paradigmatic example of vulnerability-driven coalition-building in climate governance [21]. Formed in 1990 as an intergovernmental organization comprising 39 full members and 4 observers—primarily low-lying atoll nations scattered across the Pacific, Indian Ocean, and Caribbean—AOSIS was born out of a collective recognition of shared existential perils [22]. The materials trace its origins to the late 1980s, when rising awareness of global warming's oceanic impacts prompted small island developing states (SIDS) to band together [23]. Unlike other coalitions formed around economic or political interests, AOSIS coalesced explicitly for "survival and sustainable development," as articulated in early declarations. Its membership, drawn from regions like the Pacific (e.g., Tuvalu, Kiribati) and Caribbean (e.g., Barbados, Antigua), reflects a geographic dispersion that belies a unified vulnerability: these nations, often mere meters above sea level, face inundation, salinization of freshwater, and intensified cyclones due to climate-induced sea-level rise and ocean warming.

The threats confronting AOSIS members are multifaceted and acute, as elaborated in the documents. Sea-level rise, projected to increase by 0.3 to 1 m by 2100 under moderate scenarios, poses an immediate survival risk. For instance, Tuvalu—a Polynesian archipelago with a population of around 11,000—has already initiated contingency plans for national relocation, as rising waters erode coastlines and contaminate aquifers. Ecosystem destruction compounds this: coral bleaching, driven by ocean acidification and warming, devastates marine biodiversity, undermining fisheries that sustain local economies. Dry spells and altered precipitation patterns exacerbate water scarcity, while socioeconomic instability manifests in forced migrations, health crises from vector-borne diseases, and economic losses from tourism disruptions. The materials reference UN data indicating that SIDS, despite emitting less than 1% of global greenhouse gases, suffer annual losses equivalent to 2–3% of GDP from climate events—figures that dwarf those of larger economies.

Historically, AOSIS's development mirrors the broader arc of climate negotiations. In the early 1990s, during the Rio Earth Summit preparations, AOSIS advocated for the inclusion of sea-level rise in the UNFCCC text, pushing for special recognition of SIDS vulnerabilities in the Barbados Programme of Action (1994). This marked its transition from ad hoc advocacy to institutionalized presence. By the Kyoto Protocol era (1997–2005), AOSIS intensified calls for Annex I countries to reduce CO2 emissions by at least 20% below 1990 levels by 2005, emphasizing Northern historical responsibility while urging all nations, including developing ones, to participate in mitigation. The 2009 Copenhagen Conference represented a pivotal escalation: AOSIS championed the 1.5 °C limit, contrasting with the 2 °C favored by major emitters, and threatened walkouts if Kyoto was undermined. Although Copenhagen yielded a non-binding accord, AOSIS's moral suasion laid groundwork for future gains.

The Paris Agreement (2015) epitomized AOSIS's agenda migration success. Through the High Ambition Coalition—a cross-camp alliance with the EU, LDCs, and select Latin American states—AOSIS secured the 1.5 °C aspiration in Article 2, alongside provisions for enhanced adaptation finance and loss and damage mechanisms. Post-Paris, AOSIS has sustained momentum, advocating at COP27 for operationalizing the Loss and Damage Fund and integrating ocean-based solutions like blue carbon sinks. The materials highlight AOSIS's strategies: moral diplomacy, framing climate as a "human rights" issue; cross-alliances, transcending North–South divides; and knowledge mobilization, partnering with IPCC and NGOs to amplify scientific evidence of vulnerability.

Challenges persist, as the documents underscore. Resource constraints—limited populations, fiscal capacities, and technical expertise—hamper implementation. Adaptation costs, estimated at billions annually for infrastructure like sea walls and resilient agriculture, strain budgets reliant on tourism and aid. International cooperation hurdles include political inertia from major emitters and complex funding access under mechanisms like the Green Climate Fund. Prospects, however, are politics-dependent: stable global relations could bolster support, while advancements in climate science (e.g., refined sea-level models) and technology (e.g., renewable microgrids) offer pathways forward. AOSIS's ongoing advocacy—issuing urgent warnings at forums like the 2023 Dubai Summit—positions it as a sentinel for global action, reminding the world that "our islands are sinking" not as rhetoric, but as reality.

11.4.2 LDCs: Appeals and Challenges

The Least Developed Countries (LDCs) group, comprising 46 nations primarily in sub-Saharan Africa, South Asia, and the Pacific, represents another facet of vulnerability politics [24]. Defined by the UN based on low income, human asset weakness, and economic vulnerability, LDCs host approximately 40% of the world's poorest populations—over a billion people—yet account for less than 4% of global emissions. The materials emphasize their paradoxical plight: minimal historical "contributions" to climate change juxtaposed against maximal contemporary impacts, including droughts, floods, and agricultural collapses that exacerbate poverty cycles.

LDCs' appeals in negotiations are rooted in principles of equity and survival. On mitigation, they demand ambitious Northern reductions aligned with 1.5 °C pathways. For example, at COP26, they proposed the US cut emissions 195% below 2005 levels by 2030 (70% domestically, 125% via $80 billion annual aid), the EU 65% below 1990 with $33–36 billion financing, and similar targets for the UK, Canada, Australia, and Japan. Adaptation appeals focus on establishing a Global Goal on Adaptation (GGA), streamlining funding access, and mandating 50% of climate finance for adaptation via channels like the Green Climate Fund [25]. Loss and damage, formalized in Paris Article 8, sees LDCs pushing for the Santiago Network on Loss and Damage (SNLD) operationalization, including technical assistance linking needy states with providers. Finance scaling post-2025 target's needs-based goals, addressing the 2009 Copenhagen pledge of $100 billion annually, which remains unfulfilled—actual flows estimated at $22 billion after adjustments for overreporting.

Challenges are manifold, as detailed in the documents. Funding mismatches plague delivery: OECD reports $71.2 billion in 2017, but analyses reveal only $22 billion in true value due to loans, interest, and double-counting. Loan dominance—80% of public finance as non-grants—burdens debt-vulnerable LDCs, contradicting UNFCCC principles of additional, grant-based aid. Access barriers include cumbersome applications and lack of accountability in reporting, with no agreed definitions for "new and additional" finance. Internal LDCs diversities—e.g., landlocked African states vs. island LDCs—complicate unified positions, while capacity gaps in negotiation expertise hinder effective advocacy. Despite these, LDCs emphasize Northern leadership for justice, invoking Mandela's negotiation ethos: solutions emerge from goodwill. Their strategies mirror AOSIS: victim narratives highlight "innocent victims" bearing crisis costs; institutional innovations like GGA tie adaptation to mitigation; cross-alliances with AOSIS and progressives amplify voices. Prospects hinge on fulfilling Paris commitments, with materials noting Covid-19's exacerbation of debts, urging SDRs and relief for green recoveries.

11.4.3 Shared Logic: Globalizing Climate Justice

AOSIS and LDCs share a logic of globalizing climate justice, transcending vulnerability to forge ethical imperatives. Both employ victim narratives—"our survival is at stake"—to shift climate from technical environmentalism to moral humanism, amplifying weak voices in power asymmetries. Institutional innovations abound: AOSIS's 1.5 °C push and LDCs' loss/damage advocacy have embedded equity in regimes. Cross-alliances, like High Ambition, break binaries, fostering hybrid coalitions. This shared approach has globalized justice, influencing norms like "just transition" and positioning vulnerability as a governance catalyst. Materials affirm their role in ethicizing debates, ensuring climate governance addresses not just emissions, but human dignity.

11.4.4 Group Interactions and Institutional Futures: From Confrontation to Co-Governance

The interactive dynamics among climate coalitions form the crucible of global governance, where confrontations yield to complicities, and institutional logics evolve toward futures of fragmentation, networking, and agenda competition. Building on prior sections, this analysis synthesizes materials on group behaviors, revealing how G77 + China, Umbrella Group, OPEC, AOSIS, and LDCs interweave in a multipolar arena.

11.4.5 Institutional Logic of Group Politics

Group politics fosters a "group-centric multilateralism," where coalitions mediate state actions amid systemic inequalities [26]. Materials depict this as functional adaptation: states aggregate to overcome isolation, with G77 + China enhancing Southern bargaining, Umbrella dispersing Northern burdens, and vulnerable mobilizing morals. Constraints include coordination costs—ministerial meetings demand consensus amid heterogeneities—and fragmentation, as Paris NDCs dilute bindings. Functions: coordination streamlines positions; legitimacy bolsters representation, as AOSIS's moral claims legitimize demands.

Antagonisms reproduce power on responsibility (CBDR reinterpretations), finance ($100B gaps), technology (IP barriers). Complicities emerge in hierarchical compromises, like Loss and Damage Fund, maintaining stability through "dynamic equilibrium." Polycentric trends see non-state actors (C40 cities) erode group monopolies. Networking via issue alliances (High Ambition) transcends identities. Agenda-ization shifts to norm contests ("just transition" vs. CBDR). Coalitions drive dialectics—from Rio divides to Glasgow collaborations—missioning synergistic co-governance for justice.

11.5 Conclusion

In conclusion, the dynamics of group politics in global climate governance illuminate the profound tensions and adaptive capacities within the international system. From the G77 + China's journey from unified Southern solidarity to internal fragmentation, to the defensive maneuvers of the Umbrella Group and OPEC, and the moral leverage exercised by AOSIS and LDCs, coalitions have served as both catalysts for progress and barriers to consensus. These entities, shaped by economic disparities, ecological vulnerabilities, and ethical imperatives, have redefined the North–South divide into a multipolar landscape, where principles like CBDR evolve amid shifting power balances [27]. As evidenced from Rio to Glasgow, group interactions—marked by antagonisms over responsibility and finance, yet yielding complicities in hybrid alliances—underscore the need for institutional reforms that foster inclusive co-governance.

Looking ahead, the future of these coalitions hinges on navigating polycentric trends, where non-state actors and networked agendas increasingly influence outcomes. By addressing internal heterogeneities and external pressures, such as unfulfilled funding pledges and technological barriers, these groups can transition from confrontation to collaborative stewardship. Ultimately, their evolution reflects the broader Anthropocene challenge: forging equitable solutions in a fragmented world, ensuring that climate justice transcends rhetoric to underpin sustainable planetary futures [28].

References

1. Yu H, Wang J (2012) Global climate governance and developing countries' climate negotiation strategies research. Mao Zedong Deng Xiaoping Theory Res 12:70–76
2. Zhuang Y, Chen Y (2014) Analysis of state groups in international climate negotiations and their influences. Pac J 22(12):68–78
3. Cullet P (1999) Differential treatment in international law: towards a new paradigm of inter-state relations. Eur J Int Law 10(3):549–582
4. Wu J et al (2013) Analysis of state groups in international climate negotiations. Bull Chin Acad Sci 28(6)
5. Lanlan X (2009) The "G77+China" in the climate field [J]. World Environ 06:42–44
6. Sun X, Li Y (2013) Research on the climate change cooperation mechanism between China and the Group of 77. Int Political Stud 34(1):88–102, 2, 1
7. Betzold C, Castro P, Weiler F (2012) AOSIS in the UNFCCC negotiations: from Unity to fragmentation? Clim Policy 12(5):591–613. https://doi.org/10.1080/14693062.2012.692205

8. Qimin C, Chuan T, Xiang G, Xu, Huaqing (2015) Comparative study on the BASIC countries' cooperation mechanism and low-carbon development models [J]. Comp Econ Soc Syst 03:106–114
9. Wan P (2014) Analysis of the positions of BASIC countries in international climate negotiations. Master's thesis, China Foreign Affairs University
10. Shuangwu Y, Lanlan X (2010) Divisions between China and the G77 in international climate negotiations [J]. Contemp Int RelatS 04:21–26
11. Falkner R (2019) The Paris Agreement and the new logic of international climate politics. Int Aff 95(5):1107–1125
12. Zhu S (2017) From Copenhagen to Paris: the evolution and development of the international climate regime, 2017th edn. Tsinghua University Press, Beijing
13. Xu X (2016) Changes in developing countries' group relations in international climate negotiations and their impacts. Doctoral dissertation, Shandong University
14. Li C (2010) Climate change and climate justice. J Fuzhou Univ (Philos Soc Sci Ed) 6
15. Huang P (2021) COP observations: viewing international climate political games from COP26. Greater Bay Area Review. 9 Nov 2021
16. Gao X (2010) Analysis of the positions of umbrella group countries in post-Kyoto climate negotiations. Int Forum 12(4)
17. Zhang Q, Wang C, Wang D (2014) The impact of umbrella group's climate negotiation positions on china's participation in international climate negotiations. Product Res 10
18. Li X, Xu Y, Xiao X (2022) Evolution trends of international oil trade patterns and china's responses in the context of the russia-ukraine conflict. Int Econ Coop 3
19. Zhang L (2020) OPEC's public diplomacy in the context of global climate governance. Public Diplom Q 1
20. Chen J (2020) Research on Islamic countries' participation in global governance. Doctoral dissertation, Shanghai International Studies University
21. Yan N (2012) Alliance of small island states in international climate negotiations. Master's thesis, China Foreign Affairs University
22. Pei W, Qi D (2010) Alliance of Small Island States: coping with matters of life and death [J]. Outlook 14:38–39
23. Yaman C (2011) Alliance of Small Island States: where is our future? Int Rev 06:14–15
24. Adger N, Vincent K (2005) Uncertainty in adaptive capacity. CR Geosci 337:399–410
25. Gyeltshen N (2021) COP26 conference: least developed countries need funding to support climate adaptation. China Dialogue. 28 Oct 2021
26. Liu H (2019) State group politics and international institutional competition: taking climate governance as an example. World Econ Politics 6
27. Keohane R, Victor D (2016) Cooperation and discord in global climate policy. Nat Clim Chang 6:570–575
28. Held D, Roger C (2018) Climate governance in the anthropocene: institutions and legitimacy. Polity Press, Cambridge
29. Zhang X (2014) Internal divisions in the G77 group in post-Copenhagen climate negotiations and China's countermeasures. Master's thesis [or Doctoral dissertation], University of International Business and Economics
30. Lü X (2012) Analysis of the principles of responsibility in responding to climate change in the framework convention on climate change. J China Univ Political Sci Law 3

Chapter 12
Non-state Actors: Pushing for Change from the Ground up

The process of international climate change law is increasingly emphasizing the role of non-state actors. Particularly, the Paris Agreement adopted a novel model compared to the Kyoto Protocol, moving away from mandatory emission reduction requirements towards procedures and mechanisms that encourage parties to voluntarily propose and enhance their climate targets, facilitating synergistic policy adjustments. This shift is widely regarded as a transition from a "regulatory" to a "catalytic and facilitative" model, representing an innovative response to the challenges faced by the Kyoto Protocol and the failure of the Copenhagen climate negotiations. Against this backdrop, the role of sub-national and non-state actors in international climate action has become increasingly prominent. The Decision adopted by the Conference of the Parties to the Paris Agreement introduced the concept of "non-party stakeholders," explicitly acknowledging the proactive role of non-state actors—such as non-governmental organizations, cities, and multinational corporations—in global climate governance [1].

Academic perspectives on the nature and definition of non-state actors vary, leading to different typologies. Some scholars, drawing on the geographical concept of scale, categorize them into three groups: super-national actors (e.g., international intergovernmental organizations, regional state alliances), transnational actors (e.g., international NGOs, multinational corporations, transnational violent groups), and sub-national actors (e.g., local NGOs, businesses, individuals) [2]. Others classify them based on their inherent nature: sub-national governmental authorities like states, cities, and regions; market-based actors such as corporations and business alliances; and social actors including research institutions and NGOs—a typology notably proposed by Chinese scholar Li Xinlei [3]. Reflecting the practical involvement of non-state actors in global climate governance, this chapter focuses its examination on several key categories: sub-national actors like cities and regions; academic institutions such as research bodies and think tanks; non-governmental organizations; and multinational corporations.

X. Zou, *Thirty Years of Global Climate Governance*, Energy, Climate Change & Sustainability in the Global South, https://doi.org/10.1007/978-981-95-8457-4_12

12.1 Multinational Corporations in Global Climate Governance

As the severity of global climate change becomes ever more apparent, climate governance has emerged as a shared challenge for the global community. In this process, multinational corporations (MNCs), as core participants in the global economy, play an indispensable role in promoting emission reduction and sustainable development, leveraging their extensive international influence, substantial technological prowess, and capital advantages. However, corporate engagement is not a one-sided contribution; MNCs themselves face numerous governance challenges and practical dilemmas. An objective appraisal of their dual role is essential to better guide their constructive participation in global climate governance [4].

12.1.1 The Positive Role of MNCs in Climate Governance

Multinational corporations are diverse, but can be broadly categorized by their business nature into resource extraction, manufacturing, and service provision. Despite operating in different sectors, their approaches to participating in climate governance share commonalities, primarily manifesting through three pathways: proactively seeking cooperation with intergovernmental organizations like the UN; actively participating in climate initiatives led by international NGOs; and acting as independent entities by autonomously setting and implementing climate targets to fulfill corporate social responsibility [5]. Regardless of the approach, MNCs' participation often yields tangible or long-term benefits. Owing to their operational flexibility, broad societal influence, and significant financial and technological capabilities, multinational corporations demonstrate multifaceted positive impacts in climate governance [6].

First, MNCs play a demonstrative and leadership role in emission reduction actions. A cohort of forward-looking corporations has integrated the concept of "sustainable development" into their strategic core, being among the first to publicly announce and implement emission reduction targets. For instance, the international financial giant HSBC Group has significantly reduced its carbon footprint through a series of measures, including comprehensive energy efficiency improvements and procurement of green electricity, setting a benchmark for its industry and beyond.

Second, MNCs effectively drive the low-carbon transition in developing countries. Faced with dual constraints of capital shortage and technological lag, many developing nations struggle to address climate change. Through foreign direct investment, technology transfer, and collaborative R&D, MNCs introduce advanced energy-saving technologies, low-carbon products, and management models to host countries. This not only enhances local energy efficiency but also provides crucial support for building local low-carbon economic systems.

Furthermore, MNCs are pivotal forces in the development and dissemination of new energy technologies. To align with the global energy transition trend, numerous corporations have increased investments in renewable energy, clean fuels, and related sectors. For example, Ford Motor Company significantly expanded its bio-ethanol fuel infrastructure and increased production of hybrid electric vehicles tenfold, actively promoting carbon reduction in the transportation sector.

Finally, MNCs have contributed to the refinement of global environmental governance mechanisms. In June 1993, the International Organization for Standardization established the Technical Committee for Environmental Management (ISO/TC 207), responsible for developing the ISO 14000 series of international environmental management standards. In the year these standards were published, 1,491 enterprises and organizations worldwide obtained ISO 14001 certification; by the end of June 2001, this number exceeded 30,181 [7]. Thus, the support and voluntary implementation of international environmental standards by MNCs have also played a role in advancing global environmental governance frameworks.

12.1.2 The Negative Role of MNCs in Climate Governance

Nevertheless, it is crucial to soberly recognize that the immense scale and inherent profit-seeking nature of multinational corporations also render them a source of severe challenges within climate governance. Their negative impacts are primarily manifested in the structural lock-in of carbon emissions, overconsumption of resources, exacerbation of social inequities, transboundary transfer of pollution, and resistance to robust climate policies.

The primary and most fundamental challenge lies in the fact that many MNCs are themselves massive emitters of greenhouse gases. With operations spanning the globe and often deeply embedded in energy-intensive industries such as fossil fuels, heavy chemicals, and aviation—or reliant on globalized, elongated, high-carbon supply chains—their aggregate carbon footprint is staggering. Their existing asset structures and business models create a form of "lock-in effect" for high-carbon development pathways, presenting an objective impediment to rapid global decarbonization.

Secondly, to sustain their global production networks, MNCs' demand for natural resources is voracious. Whether through establishing facilities in resource-rich areas or extracting vast quantities of water for products like bottled beverages and fast-moving consumer goods, their global resource extraction and utilization behaviors frequently lead to local resource depletion, biodiversity loss, and ecosystem degradation. This, in turn, undermines the capacity of natural systems to buffer and adapt to climate change.

On a social dimension, some MNCs, while operating in developing countries, may exacerbate resource inequality and socio-economic exploitation through issues such as land acquisition, water resource competition, and insufficient protection of labor rights, sometimes even triggering community conflicts. Such unsustainable operational practices not only harm social well-being but also erode the social stability essential for successful climate action.

More insidious yet profoundly harmful is the phenomenon illuminated by the "Pollution Haven Hypothesis": some enterprises may exploit disparities in national environmental regulatory standards by relocating heavily polluting production processes to countries with lower environmental standards. This results in the local discharge of hazardous waste, emissions, and wastewater, causing transboundary pollution [8]. This behavior violates environmental justice and effectively transfers the costs of climate governance to the most vulnerable nations and populations.

Finally, motivated by the preservation of short-term commercial interests, some MNCs heavily reliant on fossil fuels or high-carbon products, along with their industry associations, engage in organized and systematic political lobbying. They pressure governments to weaken, delay, or obstruct the introduction and implementation of critical climate policies. This "anti-climate lobbying" erodes the ambition and effectiveness of global climate governance at its policy source, representing a stubborn obstacle to global coordinated action.

Despite these undeniable negative impacts, it is imperative to recognize the immense transformative potential inherent within multinational corporations. Guiding their transition from being part of the problem to becoming central to the solution requires a multi-pronged approach: corporate self-awakening to integrate long-term climate security into core business strategy, governments establishing effective "carrot-and-stick" policy and regulatory frameworks [9], and sustained oversight from the public and civil society. Only through such a polycentric governance model can the power of MNCs be steered onto a trajectory aligned with global climate goals, ultimately making them constructive partners in building a net-zero future.

12.2 Case Study: Royal Dutch Shell

Royal Dutch Shell, a behemoth in the global energy industry, operates a business network spanning approximately 140 countries, holding a pivotal position in the global energy supply chain. Shell's business structure is primarily divided into four segments: Upstream, focusing on oil and gas exploration and development; Integrated Gas and New Energies, dedicated to exploring low-carbon transition pathways; Downstream, responsible for refining crude oil into various petroleum and chemical products to meet global energy demand; and Projects & Technology, concentrating on investment in low-cost, innovative projects and R&D. This diversified portfolio reflects both its traditional fossil fuel-dominated foundation and strategic foresight in responding to the energy transition.

Historically, the international oil industry witnessed several major environmental events and policy shifts between 1960 and 2021, with the industry's stance on climate change gradually evolving from passive response to active engagement. As a quintessential representative of multinational oil corporations, Shell's climate governance behavior is not only crucial for its corporate image but also serves as a bellwether for the pace of the entire energy sector's transformation. In recent years, Shell has frequently made public statements on climate issues, unveiling a series of emission reduction targets and transition pledges, demonstrating strategic adjustments by a traditional energy enterprise amid the wave of global climate governance.

12.2.1 Shell's Climate Ambition

Climate Plan: The 2050 Net-Zero Emission Blueprint

Shell has explicitly committed to becoming a net-zero emissions energy business by 2050 or sooner, aligning with the Paris Agreement's global temperature goal. This target encompasses not only direct emissions from its own operations (Scope 1) and indirect emissions from purchased energy (Scope 2) but also extends to emissions generated during the use of its sold energy products (Scope 3), demonstrating a life-cycle carbon management ambition. The core pathway to achieving this goal involves systematically reducing carbon intensity and net carbon footprint, fully integrating carbon constraints into strategic decision-making and resource allocation. Shell defines a net-zero emissions energy business as one that does not add to the total volume of greenhouse gases in the atmosphere. The central methodology for achieving this ambition is the focus on carbon intensity and net carbon footprint.

Phased Climate Targets

As a significant greenhouse gas emitter, oil and gas firms have a responsibility to disclose relevant carbon emission information and climate-related corporate risks while actively undertaking actions to address climate change. Shell's plan includes three key objectives:

Addressing Scope 1 and Scope 2 emissions.

Reducing the carbon footprint of its products: Shell aims to lower the carbon intensity of its products by shifting its energy mix towards greater use of renewables, biomass, hydrogen, etc.

Adjusting medium to long-term carbon footprint reduction plans: In response to the heightened Paris Agreement ambition to limit warming to 1.5 °C, Shell revised its previous target of a 50% carbon footprint reduction by 2050 to 65%, alongside achieving a 30% reduction by 2035.

Drivers Behind Climate Governance Goals: A Triad of Policy, Technology, and Market

The formulation of Shell's proactive climate governance targets stems from combined pressures stemming from a changing external environment, primarily reflected in three key dimensions: strengthening policy regulations, accelerating technological change, and shifting market preferences. These three driving forces are intertwined, collectively pushing this traditional energy giant to reposition its development strategy.

The global climate governance framework has progressively matured, leading to continuously tightening policy constraints on oil and gas companies. The scientific basis provided by the IPCC's Fifth Assessment Report, confirming the anthropogenic cause of climate change, has underpinned stronger climate policies worldwide. Former Shell CEO Ben van Beurden publicly stated that the oil and gas industry faces an existential threat if it fails to proactively transform, while emphasizing the positive correlation between low-carbon development and corporate financial success [10]. This recognition occurs against a backdrop of increasingly stringent international climate agreements, Nationally Determined Contributions, and policy instruments like carbon border adjustment mechanisms, compelling energy firms to integrate climate risk into core strategic considerations.

Technological innovation has provided feasible pathways for the energy transition, also serving as a key driver for Shell's climate targets. Recent years have seen significant cost reductions in renewable power generation, breakthroughs in energy storage, and accelerated commercialization of low-carbon energy sources like hydrogen and bioenergy. These advancements are not only altering the competitive foundations of the energy sector but also reshaping the market landscape. Shell, as a traditional energy company, recognizes that maintaining future competitiveness necessitates technological innovation and a low-carbon transition. The company is investing in R&D for renewables, Carbon Capture, Utilization and Storage (CCUS), and other technologies, seeking to reconfigure its predominantly fossil-based energy structure and find new avenues for sustainable development.

Shifts in market forces are equally significant. Rising public environmental awareness has strengthened consumer preference for green products and services. Investors are increasingly focused on corporate Environmental, Social, and Governance (ESG) performance, and supply chain partners are setting carbon neutrality requirements. This shift in market preferences influences Shell's decisions through multiple channels: consumers lean towards low-carbon energy products, while financial institutions and investors' aversion to high-carbon assets increases the company's financing costs and reputational risk [11]. Shell needs clear climate targets and transparent progress disclosure to respond to market expectations for a green transition, thereby safeguarding its brand image and market position.

Driven by this triad of forces, Shell's climate governance targets represent both a response to external pressures and a strategic choice based on long-term corporate development. Policy regulation sets the baseline for transformation, technological progress provides the implementation path, and market preferences determine the commercial value. Together, they compel Shell to transform climate governance from a reactive posture to proactive planning, driving the company's transition towards becoming a low-carbon energy provider.

12.2.2 Shell's Climate Governance Measures

Shell's role in global climate governance is multifaceted. It positions itself not only as an energy provider but also as a practitioner of the energy transition and a facilitator of sectoral collaboration. Specifically, Shell participates in climate governance through three complementary roles: as an energy provider, advancing the decarbonization of its own energy products and influencing the broader market; as an energy user, systematically optimizing the efficiency of its own operations and reducing carbon intensity across its value chain; and as a partner in decarbonizing energy use, actively providing low-carbon solutions to help its customers reduce their emissions.

Regarding process improvement and energy efficiency, Shell has established a systematic carbon management system. The company mandates that operating assets with annual emissions exceeding 50,000 tonnes must develop detailed Greenhouse Gas Management Plans, complemented by energy efficiency improvement schemes. For example, measures such as optimizing refining processes, upgrading key equipment, and applying digital monitoring systems have effectively reduced production-phase carbon intensity. Concurrently, Shell has developed tailored emission reduction solutions for customers in different sectors, promoting high-efficiency lubricants and low-carbon fuels in transport, and offering energy management consultancy and process optimization services in industry.

In optimizing its energy mix, Shell is steadily advancing its "less oil, more gas" strategy, explicitly targeting an increase in the share of natural gas in its portfolio to over 55% by 2030 and 75% by 2050. This strategic pivot reflects the company's view of gas as a transition fuel, while it also actively invests in new energy ventures. Through investments in wind, solar, hydrogen, and other projects, Shell is gradually expanding its low-carbon energy portfolio, for instance, participating in offshore wind development in Europe and North America, and promoting hydrogen infrastructure in Asian markets.

In carbon technology R&D and ecological compensation, Shell views CCUS as a critical pathway for deep decarbonization. Building on the operational success of the Quest project in Canada—which captured and stored approximately 5.5 million tonnes of CO2 between its 2015 startup and the end of 2020 [12]—Shell plans to build 25 similar-scale CCS facilities by 2035, adding 25 million tonnes of annual storage capacity. Furthermore, the company enhances carbon sink capacity through large-scale afforestation, wetland protection, and other Natural Climate Solutions. For instance, Shell's forest conservation projects in Indonesia and Brazil achieve carbon compensation goals while delivering co-benefits for local ecosystems and community development.

The systematic implementation of these measures reflects Shell's holistic approach as a traditional energy company to addressing climate change: optimizing operational efficiency through internal governance, reshaping its business model through energy mix restructuring, and breaking through emission reduction bottlenecks via technological innovation, ultimately aiming to align corporate strategy with global climate objectives. However, it is noteworthy that despite significant investments in areas like CCUS, the proportion of total investment allocated to new energy businesses remains relatively limited, highlighting the challenge of balancing transition pace with practical constraints.

Effectiveness and Limitations of Shell's Participation

Shell has achieved phased results in its climate response, reflected in quantifiable metrics. Using net carbon footprint as a key indicator, Shell reduced this value to 75 g of CO2 equivalent per megajoule in 2020, a 4% decrease from 2019 and a cumulative 5% reduction since 2016. This progress indicates sustained efforts in energy efficiency and optimizing its energy mix. In carbon technology practice, Shell's Quest CCUS project has demonstrated significant efficacy since its 2015 launch, successfully capturing and storing about 5.5 million tonnes of CO2 by end-2020, serving as an important example of industrial-scale CCUS application [12].

However, as a traditional oil and gas enterprise, Shell faces structural limitations. Its core business remains dominated by fossil fuels, and its operational boundary targets do not fully encompass Scope 3 emissions (product use), inherently limiting its potential contribution to global climate governance. The landmark 2021 ruling by the District Court in The Hague against Shell underscored this contradiction. The court found Shell's climate policy insufficient and ordered it to accelerate its emissions cuts [13]. This case reflects growing public demand for corporate climate accountability and exposes challenges such as insufficient legal constraints and ambiguous responsibility definitions for multinationals in climate governance.

Responding to various pressures, Shell accelerated its strategic pivot, launching new strategies like "Powering Progress" to balance business interests with climate responsibilities through portfolio adjustments and innovation. Yet, BloombergNEF statistics indicated that as of June 2023, Shell's substantive investments in low-carbon areas remained modest. Among globally retired carbon credits, about 20% were used by oil and gas firms, with Shell retiring over 100 million tonnes of CO2 equivalent, primarily from avoided deforestation projects. In contrast, its capital expenditure in solar, wind, hydrogen, and other low-carbon energy sources accounted for only 9% of total investment [14]. This data折射 (highlights) the transition dilemma facing Shell and the wider oil and gas industry: reliance on carbon offsetting for short-term targets, coupled with conservative investment in disruptive renewable technologies.

The case of Shell demonstrates that traditional energy giants possess advantages in capital, technology, and management experience for climate governance, yet are constrained by path dependency and their business structures. Looking ahead, achieving a virtuous cycle between climate governance and corporate development requires not only enhanced innovation and strategic renewal by companies themselves but also synergistic advancement of policy guidance, market mechanisms, and public oversight [15]. Only through multi-stakeholder collaboration can energy firms like Shell be genuinely propelled towards a low-carbon, sustainable future.

12.3 Think Tanks in Global Climate Governance

Think Tanks, emerging as crucial knowledge producers and policy advisory bodies in response to the increasing complexity and scientization of modern state decision-making, have seen their policy influence grow substantially since the twentieth century, particularly against the backdrop of globalization and the rising prominence of climate change. In the realm of climate governance, think tanks not only deeply engage in scientific assessment, policy research, and transnational dialogue but also act as key bridges connecting science, policy, and the public, propelling global climate governance from conceptual framing towards concrete action.

Think tanks cover a wide range of research fields, with a significant number focusing on environmental protection and energy. As climate change has garnered escalating attention within societies worldwide due to globalization, it has similarly captured the focus of think tanks. Taking U.S. think tanks as an example, the Brookings Institution, among the most prestigious comprehensive think tanks, is often considered the origin of the modern think tank [16]. As the international context evolved, Brookings shifted its focus beyond domestic issues to increasingly engage in regional policy studies, global development, urban governance, and more. Peter J. Wilcoxen and Warwick J. McKibbin were among the earliest researchers at Brookings to venture into environmental and climate issues [17]. Entering the twenty-first century, climate research at Brookings gradually increased, peaking around 2009 [18]. Its policy prescriptions emphasized that the U.S. government should prioritize climate change, enhance technological innovation and application, accelerate the process of emissions trading, and strengthen international cooperation.

Beyond comprehensive think tanks, specialized think tanks focusing on environment and climate play an irreplaceable role in global governance. The World Resources Institute (WRI), a leading global environmental policy research organization, has formed a systematic research network covering climate, energy, forests, water, and other areas. By establishing offices in countries like China, Brazil, and Indonesia, WRI promotes transnational knowledge sharing and localized policy innovation. Other highly influential institutions, such as the Center for Climate and Energy Solutions (C2ES), Resources for the Future (RFF), and the Center for Energy and Environmental Policy Research (CEEP), provide scientific basis for national emission reduction strategies through technical pathway analysis, policy evaluation, and scenario modeling. The annual *State of the World* report published by the Worldwatch Institute has become a key reference in the field of global environmental governance.

These think tanks participate directly in the climate governance process not only through research reports, policy briefs, and technical solutions but also by actively constructing transnational cooperative networks, fostering a shared but differentiated understanding of responsibility. Think tanks also contribute to the transparency and accountability of global climate governance by participating in UN climate change conferences (COPs) and publishing climate performance indices. It is noteworthy that think tanks also exhibit differences in stance and methodology within climate governance. Some Western think tanks, influenced by funding sources and political agendas, may produce policy recommendations that deviate from impartial positions, occasionally exhibiting tendencies towards "strategic research." Furthermore, in promoting "climate communication," issues such as a West-centric discourse system and disconnects between proposed solutions and the actual needs of developing countries can arise.

Despite these challenges, think tanks have demonstrated considerable adaptability and innovation in global climate governance. In recent years, think tanks from emerging economies like China, India, and Brazil have developed rapidly, gradually breaking the Western monopoly on climate discourse. For instance, Chinese think tanks, through mechanisms like the Belt and Road Initiative for green and low-carbon cooperation and South-South climate cooperation, promote technological pathways and governance solutions more suited to developing countries. Against the backdrop of global carbon neutrality goals, think tanks will further collaborate with governments, businesses, city networks, and other actors, solidifying their role as essential forces in building a fair and effective global climate governance system.

12.3.1 Case Study: World Resources Institute (WRI)

The World Resources Institute (WRI), founded in 1982 by prominent environmental advocate James Gustave Speth and headquartered in Washington, D.C., is a specialized environmental and climate think tank. Grounded in scientific research, WRI is dedicated to fostering synergistic development between society, the economy, and the environment. By providing practical policy solutions and technical tools, it aims to help achieve the dual goals of protecting the planet and enhancing human well-being.

According to its official information and public data, WRI has established offices in over 12 countries and regions globally, employs more than 1,700 experts and staff, and its projects extend to more than 50 countries. In 2008, WRI opened its first overseas office in Beijing, marking its formal inclusion of China into its global strategic network and reflecting the high importance it places on China's role in global environmental governance. WRI's research framework systematically revolves around seven core themes: Climate, Energy, Food, Forests, Water, Ocean, and Sustainable Cities. Methodologically, WRI emphasizes a combination of "quantitative analysis—local practice—global scaling," using data-driven research, pilot projects, and capacity building to drive policy formulation and innovation in implementation pathways. For example, its Aqueduct Water Risk Atlas tool and its co-leadership of the Science Based Targets initiative (SBTi) have been widely adopted as key environmental management tools across multiple countries and sectors.

As a leading global think tank on environment and development, WRI systematically focuses its research activities on seven critical areas, with climate change consistently at its core. Leveraging objective, data-driven research, WRI has exerted significant influence in shaping and informing the debate on U.S. climate policy. Its research outcomes not only provide crucial evidence for policymakers but also help shape public and business understanding of the necessity for climate action. This evidence-based, in-depth engagement has earned WRI outstanding academic credibility and broad policy influence, both within the United States and internationally.

a) **WRI's Research Focus and Key Viewpoints**

Research Areas and Programs

Since 2013, WRI has embarked on a series of strategic innovations, explicitly focusing its future research on seven critical challenges facing global sustainable development. Around these seven core issues, WRI conducts multidimensional, cross-sectoral systematic research and practical projects aimed at driving tangible change through scientific assessment and policy advocacy.

- In **Climate**, WRI is committed to assisting local governments, national institutions, businesses, and civil society in collaboratively promoting the deep economic and social structural transformations needed to address climate change and implement global GHG emission reduction pathways.
- In **Energy**, through its "Global Energy Program," WRI actively promotes innovation in clean energy market mechanisms and institutional capacity building, guiding more investment towards clean energy technologies that support community prosperity and public health.
- Regarding **Food** systems, WRI focuses on global food loss and waste, analyzes sustainable yield enhancement strategies, and develops environmentally friendly agricultural production methods.
- On **Forests**, WRI collaborates with governments, businesses, and NGOs to curb illegal deforestation, promote restoration of degraded lands, and maintain critical forest ecosystems.
- Concerning **Water**, WRI addresses global water security challenges through multi-stakeholder cooperation, tackling water scarcity and pollution.
- In **Sustainable Cities,** WRI emphasizes balancing economic vitality with ecological protection, collaborating with governments in China, India, Brazil, Mexico, and Turkey on innovative governance projects for rapid urbanization.

On **Ocean** protection, WRI focuses on marine ecosystem degradation and explores sustainable management pathways for marine resources.

Key Viewpoints

WRI consistently positions climate change as an urgent, systemic threat to global society and ecosystems, emphasizing that its intensifying impacts demand rapid, decisive, and coordinated international action. As a key think tank driving global climate governance, WRI's core mission is to assist decision-makers—including international organizations, national governments, businesses, and civil society—in collaboratively advancing deep structural transformations of economic and social systems to systematically address the climate challenge.

To achieve this, WRI proposes initiatives like its "International Climate Action Initiative" to strengthen multilateral cooperation and intergovernmental policy coordination, fostering more ambitious national climate policies and facilitating the implementation of the Paris Agreement goals. In engaging non-state actors, WRI launched and leads the Science-Based Targets (SBT) initiative, which has helped hundreds of companies worldwide set emission reduction targets aligned with climate science. The Greenhouse Gas Protocol (GHG Protocol), developed under its leadership, has become the international standard for businesses, cities, and even nations to measure and manage emissions, providing the foundational tool for global carbon accounting.

WRI's governance participation reflects a distinct global perspective and a propensity for multilateral cooperation. Its work aims not only to influence U.S. domestic climate policy but also to build transnational, cross-sectoral collaborative networks, fostering an inclusive and effective global climate governance framework. WRI's policy influence is fully evidenced in its series of authoritative reports. As early as 2011, its *World Resources Report 2010–2011: Decision Making in a Changing Climate* was prescient. Synthesizing research and insights from climate experts in over 30 countries, the report unequivocally identified effective government decision-making as the core of responding to the climate crisis and systematically proposed five key elements for effective decisions: (1) Broad public participation; (2) Solid data support; (3) Sound institutional mechanisms; (4) Integration of diverse resources; and (5) Application of advanced tools [19]. The report profoundly stated that climate change is not just an environmental challenge but a severe test of a government's governance capacity, requiring decision-makers to make difficult and wise trade-offs between short-term interests and long-term sustainable development.

b) **Pathways for Influencing Climate Policy**

In its engagement with global climate governance, WRI has gradually developed a multi-level, multi-channel system for policy influence. Its target audiences span the core decision-makers, the specialized intermediate layer, and the broader public [20], employing differentiated communication and interaction strategies to effectively advance the scientization and policy integration of climate issues.

The "Revolving Door" Mechanism

The "revolving door" is a core mechanism of influence for modern think tanks, particularly within the U.S. system. It refers to the two-way channel between think tank researchers and government officials: experts can enter government roles, translating research findings into policy, while former officials often join think tanks, bringing valuable experience and networks into academic research, thus enabling a continuous exchange and mutual enrichment of knowledge and power.

WRI's utilization of this mechanism exhibits two distinct characteristics. First, regarding the direction of personnel flow, direct movement from WRI into senior U.S. federal government positions is relatively less common. Instead, WRI more frequently sees former government officials, senior international organization officers, and experienced policymakers "rotate into" the institute, assuming leadership roles or senior fellowships. Secondly, WRI maintains an exceptionally active flow of talent with top universities (an "academic revolving door"), with its leaders and experts often holding professorships or deanships, and vice versa. A prime example is founder James Gustave Speth, whose career trajectory—from Chair of the Council on Environmental Quality under President Carter, to founding WRI, then becoming Administrator of the UNDP, and later Dean of the Yale School of Forestry & Environmental Studies—epitomizes the multiple rotations among government, international organizations, leading think tanks, and elite academia.

Participation in Congressional Hearings

WRI leverages participation in U.S. Congressional hearings as a core and direct channel for influencing climate and environmental policy. By testifying before House, Senate, and joint committee hearings on key issues like global climate change, energy transition, and forest conservation, WRI consistently delivers research-based findings and recommendations to policymakers. For instance, its second president, Jonathan Lash, testified before Congress on four occasions, advocating for proactive U.S. climate legislation to cut emissions, ensure energy security, and promote clean energy industries, while stressing the importance of international cooperation with major emitters like China. More recently, WRI leadership, including President and CEO Andrew Steer, have used this platform, commenting on renewable energy targets set by countries like India and China and analyzing their potential global market impacts during hearings concerning Paris climate summit outcomes. Through these testimonies, WRI effectively translates cutting-edge research into persuasive policy arguments, engages early in the legislative process, helps set the policy agenda, and provides accessible expert insight on complex scientific and economic matters.

Publishing Climate Research Findings

WRI disseminates its research findings to the public in diverse formats, including comprehensive reports, policy briefs, technical guides, working papers, and implementation manuals, forming a systematic and open knowledge ecosystem. According to WRI's official website statistics, as of 2023, it had published 1,183 publicly available reports, 395 of which—one-third of the total—were tagged with "Climate" as a core topic, underscoring the centrality of this issue. All reports are publicly accessible via its website for free online reading and download, reflecting WRI's long-term commitment to climate knowledge sharing and policy transparency.

Among its publications, the *World Resources Report* is its flagship and most influential series. Initiated in 1986, this series pioneered systematic, cross-sectoral analysis of the global environment and its relationship with socio-economic development. WRI has continued this series with relative regularity, delving into major global themes such as human health and environment, climate change, ecosystem services, sustainable cities, and food security, effectively filling information gaps on the dynamic state of global natural resources. Each report is built on rigorous evidence and extensive peer review, designed to help decision-makers identify critical issues, evaluate policy options, and drive effective solutions.

Building Global Climate Governance Networks

WRI influences global climate governance not only through solid research but also by building a multi-dimensional engagement network via multi-channel knowledge dissemination, high-frequency academic exchanges, and deep institutional collaboration, effectively facilitating interaction among climate science, policy, and practice. In knowledge dissemination, WRI actively uses digital media and online platforms to transform specialized research into accessible products. Its official website systematically showcases reports, briefs, and papers, and features blogs, news releases, and data visualizations. In fostering deep dialogue and consensus building, WRI organizes high-level international conferences, thematic workshops, and high-end think tank dialogues, creating platforms for knowledge sharing and policy exchange. For example, the "U.S.-China High-Level Dialogue on Energy and Climate Change," co-hosted by WRI, has convened multiple times, bringing together experts from governments, academia, and businesses in both countries.

Furthermore, WRI places particular emphasis on synergistic collaboration with various institutions to enhance the impact of its global initiatives. It maintains long-term strategic partnerships with international organizations like UNDP, UNEP, and the World Bank, jointly advancing international climate agreements. Simultaneously, WRI operates at the local level, working with national governments and research institutes on targeted projects. This combination of top-down and bottom-up collaboration ensures WRI's research possesses a global vision while responding to specific regional governance needs.

12.3.2 Effectiveness and Limitations of WRI's Participation

The World Resources Institute (WRI) has demonstrated significant effectiveness in its participation in global climate governance, while also facing certain limitations. In terms of achievements, WRI has provided a robust evidence base for policy-making through high-quality scientific research reports. Its research helps identify gaps in global climate action and proposes concrete solutions. At the policy influence level, WRI effectively utilizes channels like the "revolving door" mechanism and Congressional hearings to directly feed its research into the U.S. domestic climate 立法 (legislative) process. Concurrently, through deep involvement in international negotiations and cooperative projects, it promotes the implementation of agreements like the Paris Agreement and facilitates the global sharing of low-carbon transition experiences from countries like China and Brazil. Moreover, by hosting over 160 academic events, building transnational expert networks, and opening access to online databases, WRI has effectively fostered dialogue among academia, policy-makers, and the public, enhancing the transparency and pluralism of global climate governance.

However, WRI's participation also has limitations. First, its policy influence is constrained by external funding and political environments. Reliance on foundation support can raise concerns about research independence, and political shifts in major countries can weaken the traction of its advocacy. Second, despite the breadth of its research, a gap remains in practical translation. For instance, its investment recommendations for urban resilient infrastructure face financial and technical barriers in developing countries, and some policy proposals lack binding force in implementation. Third, while WRI's global network is extensive, resource allocation still tends to倾斜 (tilt towards) major emitters, with relatively limited support for Least Developed Countries and Small Island Developing States, potentially exacerbating the "North–South imbalance" in climate governance.

In summary, WRI has significantly advanced the scientization and institutionalization of climate governance through scientific empowerment, policy advocacy, and cross-sector collaboration. However, to truly achieve a just transition on a global scale, it must further overcome structural constraints and strengthen localized application and equity considerations in the future.

12.4 Non-Governmental Organizations in Global Climate Governance

Non-Governmental Organizations (NGOs), recognized as a significant concept within the UN framework, refer to social organizations that operate independently of governmental systems, possess public attributes, and undertake corresponding public functions. They are active across various domains and levels of human society, influencing public opinion and policy through multiple channels. As globalization deepens, the role of NGOs in environmental issues like climate change has become increasingly prominent, gradually establishing them as indispensable actors within the global climate governance network [21].

12.4.1 The Evolving Role of NGOs in Global Climate Governance

The trajectory of NGO participation in global climate governance reveals their evolution from a relatively marginal role to a key force within transnational advocacy networks and multilateral negotiations. Traditionally, nation-states were the core subjects of climate governance. However, as the global climate governance system evolved, states and non-state actors have interactively reshaped the manifestations and institutional structures of climate politics through frequent interaction [22]. Promoted by influential international NGOs like the World Wide Fund for Nature (WWF), NGOs have gradually assumed irreplaceable roles in agenda-setting, rule-making, local implementation, and transnational oversight.

Since the UN formally initiated the climate negotiation process in 1990, participating entities have become increasingly diverse. International NGOs, starting from a relatively peripheral position, gradually secured more institutionalized pathways for participation. Scholars often divide this process into three stages [4]:

- **Phase 1: Governance Mechanism Expansion and the Germination of Institutional Participation (1997–2007)**: Following the signing of the Kyoto Protocol in 1997, the number of peripheral governance mechanisms around the UNFCCC increased significantly. The Nairobi climate conference (COP 12) in 2006 formally confirmed that NGOs could register as observers to the Conference of the Parties (COP), granting them the right to attend and observe meetings, marking the initial inclusion of NGOs into the formal multilateral climate governance track.
- **Phase 2: Actor Diversification and Rising Influence (2008–2014)**: After the 2008 global financial crisis, progress in state-level climate negotiations slowed. Post the 2009 Copenhagen Conference, the number of NGOs registered to participate in COPs surged dramatically. Through publishing analytical reports, hosting side events, and conducting joint advocacy, they significantly influenced the negotiation agenda and the broader landscape of global climate governance [24]. Although stable, institutionalized channels for participation were not fully formed in this phase, NGOs markedly enhanced their capabilities in networking, specialization, and transnational mobilization.
- **Phase 3: Deepened Institutional Interaction and Consolidated Participation (2014-Present)**: As the climate issue continued to gain prominence globally, interactions among various non-state and sub-national actors became more frequent, forming multi-level, diverse cooperative initiative networks. The 2014 Lima-Paris Action Agenda (LPAA) provided NGOs with more institutional space for participation [25]. The 2015 Paris Agreement explicitly recognized the positive role of "non-party stakeholders" (NPS) in climate action, establishing a "bottom-up" governance model that further promoted the formalization and normalization of NGO involvement in global climate governance [26].

12.4.2 The Roles and Functions of NGOs in Climate Governance

Within the global climate governance system, NGOs exert distinct influence, different from sovereign states and intergovernmental organizations, derived from their unique organizational forms and modes of operation. This influence primarily stems from three core dimensions: extensive and flexible partnership networks in global climate governance; solid expertise and scientific research capacity; and close collaborative relationships with mass media [27]. These elements collectively constitute the "soft power" foundation of NGOs in climate governance, enabling them to play unique roles across multiple levels.

Their functional positioning in climate governance can be summarized into three core roles: serving as **raisers of climate change awareness**, translating complex scientific information into publicly accessible content through various communication channels and public education campaigns, thereby effectively enhancing societal awareness of the climate crisis; acting as **providers of climate governance information**, continuously producing high-quality analysis, policy recommendations, and data products based on specialized research teams, offering crucial knowledge support for decision-makers; and functioning as **monitors of climate protection action**, promoting the fulfillment of climate commitments and strengthening the accountability mechanisms of the governance system through independent evaluation, tracking research, and social advocacy. This functional positioning allows them to effectively compensate for the shortcomings of state-centric governance models, building critical bridges between policy formulation and implementation, and between global agendas and local action [28].

a) **Pathways and Mechanisms of Engagement**

From the perspective of specific engagement pathways, NGO actions in climate governance can be systematically categorized into "upstream" and "downstream" participation models.

Upstream Engagement: Direct Influence at the Policy Source

Upstream NGOs seek to input their expertise and value propositions into the early stages of policy formulation by providing specialized advice, participating directly in intergovernmental negotiations, and submitting policy recommendations. For instance, concepts like the co-benefits of biodiversity conservation and climate adaptation have been gradually integrated into international climate谈判 (negotiation) discussions, promoted by organizations like WWF. The effectiveness of this model often relies on the organization's research credibility, expert networks, and institutional communication channels established with decision-making bodies. This is embodied in:

Scientific Assessment & Knowledge Production: Establishing professional research teams to conduct ongoing climate-related scientific research, regularly releasing assessment reports, policy briefs, and technical guidelines, transforming complex data into actionable policy advice. **Multilateral Negotiations & Institutional Framing:** Leveraging formal observer status to deeply engage in multilateral mechanisms like UN climate conferences, injecting professional insights into negotiations through side events, position papers, and intersessional consultations, and actively participating in the development of international standards and rules.

Downstream Engagement: Policy Implementation and Action on the Ground

NGOs establish partnerships with local governments, communities, businesses, and others to conduct pilot projects, technology demonstrations, and capacity-building activities, translating macro-level policies into concrete practice. This type of engagement is typically more operational and flexible, able to provide tailored solutions responsive to local needs and accumulate scalable experience through practice. It is primarily manifested in:

Public Mobilization & Social Advocacy: Employing innovative communication strategies, multimedia content, public education campaigns, and social movements to raise awareness of climate issues across sectors, making climate science more relatable to the public and creating a favorable social atmosphere for climate action. **Local Practice & Capacity Building:** Exploring context-appropriate solutions in partnership with local governments and community organizations through pilot projects and local practice. Effectively enhancing the capacity of various stakeholders to participate in climate governance through organized training, workshops, and knowledge-sharing activities. [26]

b) **Oversight Function and System Integration**

Furthermore, NGOs serve as independent third-party institutions, promoting accountability through monitoring, evaluation, and advocacy. They track the fulfillment of climate commitments by various parties by establishing monitoring mechanisms and periodically releasing evaluation reports that reveal implementation progress and gaps, thereby pushing for greater transparency. They also strengthen accountability in global climate governance by establishing complaint mechanisms and public participation platforms, providing channels for affected groups to voice their concerns [27].

These complementary and synergistic engagement pathways collectively form a comprehensive system through which NGOs influence climate governance. At the international level, NGOs actively set and guide climate agendas, participate in and monitor multilateral negotiation processes, and utilize their transnational networks to facilitate knowledge sharing and technology transfer. Domestically, they assist countries in translating international commitments into national action through policy lobbying, scientific research, public education, and project demonstration. Through these multi-pronged approaches—knowledge production, institutional engagement, social mobilization, practical innovation, and monitoring & evaluation—NGOs play an indispensable and constructive role in advancing the global climate governance process.

12.4.3 Case Study: World Wide Fund for Nature (WWF)

The World Wide Fund for Nature (WWF), established in 1961 and headquartered in Gland, Switzerland, is one of the world's largest and most influential independent non-governmental environmental conservation organizations. As of 2025, WWF operates in over 100 countries and territories, employs approximately 5,000 full-time staff, and engages more than 5 million volunteers. It has invested nearly USD 100 billion, driving over 13,000 environmental conservation projects globally. Invited by the Chinese government in 1980 to initiate giant panda conservation work, WWF was the first international environmental NGO to enter China, formally establishing a Beijing office in 1996. It has since expanded its project areas to encompass multidimensional issues including species conservation, climate change, energy transition, and the sustainable management of freshwater and marine ecosystems. Over more than three decades of global climate governance, WWF, with its scientific solutions, multi-level cooperative networks, and innovative practical pathways, has become a paradigmatic example of NGO participation in climate governance.

a) **Climate Governance Pathways and Core Roles**

WWF's role in climate governance is evident through its systematic engagement across "global-national-local" levels, integrating policy advocacy, scientific research, and field projects into a composite governance model.

At the **international level**, WWF actively participates in global climate agenda-setting and multilateral regime building. As a long-term observer at UN climate conferences, WWF deeply engages in and influences the international climate negotiation process by releasing research reports, hosting side events, and submitting policy recommendations. For instance, during COP28, WWF urged countries to strictly adhere to the 1.5 °C warming limit and proposed that Nature-based Solutions (NbS) should form the core content of Nationality Determined Contributions (NDCs). Concurrently, WWF leads and constructs multiple transnational climate action alliances, such as the Climate Action Alliance (ACA), collectively driving the accelerated decarbonization of the global energy system by uniting businesses, cities, and research institutions.

At the **national level**, WWF focuses on translating international climate goals into domestic policy and practice. In China, WWF facilitates low-carbon transition through policy advice, technical demonstration, and capacity building. In policy advocacy, WWF has consecutively published reports like *China's Future Power Generation* and "Recommendations for China's 14th Five-Year Plan on Energy Development", providing important references for China's medium- and long-term low-carbon energy strategy. Addressing the reality that industrial energy consumption accounts for nearly 70% of China's total societal energy use, WWF launched the "Corporate Climate Action Platform," assisting high-carbon sectors like steel and cement in planning emission reduction pathways and demonstrating energy efficiency improvements and renewable energy substitution.

At the **local and public mobilization level**, WWF excels at grounding global issues into actionable individual initiatives. Since introducing the "Earth Hour" campaign to China in 2009, WWF has made it one of the country's largest public environmental participation events, attracting tens of millions of participants and effectively raising public awareness of climate change. Additionally, WWF implements community-based climate adaptation projects in ecologically fragile areas of China, such as the Southwest and the Yangtze River Basin, supporting farmers in adopting climate-smart agricultural technologies to enhance the climate resilience of vulnerable groups [29].

b) **Representative Projects and Governance Innovation**

WWF places particular emphasis on exploring scalable solutions through innovative projects, with its "Climate Solver" program and "Nature-based Solutions" practices being especially notable.

The "Climate Solver" initiative, launched by WWF in Sweden in 2008, aims to identify and support small and medium-sized enterprises (SMEs) and technologies with significant emission reduction potential. By 2025, the project had selected and supported dozens of innovative technologies in China, such as bamboo-wound composite pipes and high-efficiency flexible solar cells. Evaluations suggested that if these technologies achieved their projected market share by 2025, they could reduce emissions by over 231 million tonnes of CO2 equivalent annually. The project effectively promotes the market application of low-carbon technologies by providing technical certification, resource matching, and international exposure.

In the realm of "Nature-based Solutions," WWF promotes mangrove conservation, marine ecosystem restoration, and sustainable agriculture projects worldwide. For example, WWF's "Blue Carbon" projects in the South Pacific, involving seagrass bed and mangrove restoration, both enhance coastal climate resilience and create sustainable livelihoods for local communities. Such projects highlight WWF's integrated capacity to link ecological conservation with climate response.

c) **Analysis of Effectiveness and Limitations**

Over more than three decades of climate governance engagement, WWF has achieved notable success. Through scientific reports, policy recommendations, and multi-stakeholder cooperation, it has successfully integrated key issues like "Nature-based Solutions" and "just transition" into global and national climate agendas. In developing countries like China and India, WWF has facilitated regional low-carbon transitions through technical demonstration and capacity building. For instance, its promotion of renewable energy demonstration city projects in China provided important references for local energy planning.

However, WWF's practices also face certain limitations. Firstly, as an NGO, WWF lacks binding power over national policies; its influence heavily depends on the willingness of partner governments and the local institutional environment. Secondly, some WWF projects encounter tensions between "standardized solutions" and "localized needs" during implementation, particularly in developing countries, where balancing global objectives with pressing local issues like poverty and energy access remains challenging. Furthermore, although WWF's funding sources are diverse, its reliance on external donations means project sustainability and scalability can be affected by funding fluctuations.

The World Wide Fund for Nature, through its three-tier engagement model of global agenda advocacy, national policy empowerment, and local practical innovation, plays a pivotal role in global climate governance as a knowledge producer, partnership facilitator, and action catalyst. Its case demonstrates that NGOs can provide crucial support by compensating for governmental mechanisms, promoting multi-actor collaboration, and advancing the implementation of governance solutions. Looking ahead, as climate governance complexity increases, organizations like WWF need to further strengthen partnerships with local communities, enhance project adaptability and inclusivity, and deepen exploration of underlying issues like technological justice and the fairness of transition pathways.

12.5 Cities and Transnational Networks in Global Climate Governance

In December 2015, the adoption of the Paris Agreement at COP21 marked the beginning of a new phase in global climate governance. The Agreement explicitly set the goal of holding the global average temperature increase to well below 2 °C above pre-industrial levels and pursuing efforts to limit it to 1.5 °C, providing a clear direction for collective global action. Within this framework, a diverse array of entities—including states, sub-national governments, cities, and businesses—have put forward carbon neutrality visions and roadmaps, gradually forming a new pattern of polycentric, multi-level, and interactive governance. Cities hold a particularly unique position within this system: they are both vulnerable spaces directly impacted by climate change and critical actors in implementing mitigation and adaptation actions. The legitimacy, moral imperative, and practical pathways for city participation in global climate governance have gained growing recognition within both global discourse and institutional mechanisms. Their role has shifted from passive implementers to active architects, becoming crucial bridges linking international commitments to local practice.

The significance of cities in climate governance stems from their dual attributes of substantial influence and acute vulnerability. Currently, while cities occupy only about 2% of the world's land area, they account for approximately 70% of global greenhouse gas emissions and are home to over half of the world's population—a figure projected to rise to 60% by 2030. Climate-induced extreme weather events, such as floods, heatwaves, and sea-level rise, pose severe threats to urban infrastructure, public health, and economic and social stability [31]. For instance, research by the C40 Cities Climate Leadership Group indicates that many of its member cities face increasingly frequent climate disasters, with transport system emissions accounting for over one-third of their total emissions. This confluence of vulnerability and concentrated emissions reinforces the inherent motivation for cities to undertake climate action.

City participation in climate governance is driven not only by mitigation and adaptation needs but also relies on the cooperative platforms and governance resources provided by transnational city networks. Mechanisms such as the C40 network, ICLEI – Local Governments for Sustainability, and the Climate Alliance help cities transcend national boundaries through knowledge sharing, technology diffusion, joint advocacy, and common goal setting, forming a "bottom-up" governance force. These networks have fostered a number of exemplary city climate actions, such as

Los Angeles creating tens of thousands of jobs through renewable energy development, the development of the carbon–neutral Royal Seaport district in Stockholm, and Chengdu's participation in the C40 "Green and Thriving Neighbourhoods" program. While enhancing local capacities, these networks also influence the international negotiation process through collective advocacy, creating a "reverse shaping" effect on the global climate governance agenda.

Thus, cities are no longer merely end-points for climate policy implementation. Through transnational linkages, knowledge innovation, and institutional collaboration, they have become indispensable actors within the global climate governance network. In the "bottom-up" architecture established by the Paris Agreement, cities and their cooperative networks will continue to strengthen their functions in agenda-setting, experimental innovation, and accountability advocacy, providing grassroots momentum and implementation pathways for global climate governance.

12.5.1 The Theory of Transnational City Networks

Transnational city networks can be defined as institutionalized platforms for cooperation established through voluntary, mutually beneficial, and consultative interactions among cities, facilitated by the development of the information society under globalization. These platforms enable the sharing of experiences and exchange of best practices [31]. Cities, as the actors within these networks, can conduct information exchange and policy coordination across national boundaries, generating collective action.

Multi-level governance theory offers a lens from the perspective of the global governance system to explain transnational city networks. This theory provides a robust analytical framework for understanding the diffusion of authority in global governance: authority is no longer monopolized by sovereign states but is increasingly distributed among diverse actors such as sub-national entities, transnational organizations, and non-governmental bodies [32]. Governance power diffuses "upwards" to 'inter-state' and 'super-state' levels, such as the United Nations and the European Union, and "downwards" to sub-national actors, including cities and local governments. Cities become effective climate governance actors because they occupy an intermediate level within the global multi-level governance system. They can both implement national climate policies and actions from the top down, and provide feedback with data and practical experience from the bottom up to international and national levels, serving as crucial nodes in the climate governance process from policy to implementation.

Multi-level governance theory posits that a new set of decision-making networks has emerged in global governance. While states remain dominant in matters of sovereignty and security, sub-national actors can exert influence in policy initiation, for example, through lobbying, protests, and other means. Sub-national actors also possess distinct advantages in the realm of policy implementation. By participating in city networks, cities can acquire governance knowledge and shared experiences at lower costs, enhance their ability to attract sustainable investment, and sometimes even bypass national-level constraints to directly advance more ambitious climate actions locally. For instance, despite the U.S. announcement to withdraw from the Paris Agreement under the Trump administration, 12 U.S. member cities within C40 remained actively engaged in climate action, some playing exemplary roles within specific C40 projects and networks, providing practical experiences for other cities to draw upon.

Currently, global cooperation on climate governance among sovereign states faces difficulties, including collective action dilemmas and "free-rider" problems. Developed countries often attempt to dilute "differentiated responsibilities" in favor of common responsibility, and promised financial assistance from developed countries has not been fully delivered, making climate negotiations arduous. The establishment of transnational city networks can effectively overcome these collective action dilemmas, seeking pathways to address international problems outside intergovernmental frameworks and ensuring climate action is implemented at the local level.

12.5.2 Case Study: The Global Transnational City Network C40

Established in 2005, C40 is a global transnational city network aimed at fostering cooperation on climate governance among major cities. As a leading alliance in the climate governance arena, C40 is dedicated to strengthening collaboration among cities worldwide to reduce greenhouse gas emissions, enhance energy efficiency, and place climate issues at the heart of local policy and international agendas by advancing dialogue on climate action and environmental justice. Its members include global megacities such as London, Tokyo, New York, Sydney, Beijing, and Shanghai. From 2005 to 2022, the C40 alliance has consistently participated in international climate negotiations, established urban governance models, and become a paradigm for city collectives engaging in global climate governance by building platforms for resource sharing, organizing exchanges and policy learning, and setting membership mechanisms with awards to incentivize action. Today, C40 comprises nearly 100 of the world's leading cities, representing 25% of global GDP and over 700 million people. It has initiated more than 10,000 climate actions, standing as one of the most influential city networks.

a) **Organizational Structure**

Internal Management Mechanisms

C40's governance structure includes a Chair, a Board of Directors, and a Steering Committee. The Chair provides leadership in calling for and advocating global climate action, for instance, by convening city climate summits and conducting climate diplomacy, and is typically a mayor with city management experience. The

Board of Directors is primarily responsible for overseeing C40's management and approving the organization's budget set by the Steering Committee; some board members come from foundations that fund C40. The Steering Committee, composed of mayors from member cities on a rotating basis, manages the organization's operations, including monitoring C40 cities' actions, discussing the organizational agenda, and assessing membership qualifications.

To ensure maximum participation from member cities in climate governance, C40 has established a unique admission mechanism. Member cities must earn their membership through action, not through membership fees. In 2021, the Steering Committee passed new membership standards [33]: (1) The city has adopted a resilient and inclusive Climate Action Plan aligned with the Paris Agreement's 1.5 °C goal; (2) By 2024, the city will be implementing its Climate Action Plan and is committed to halving its collective emissions by 2030; (3) The city uses the necessary financial, regulatory, and other tools to confront the climate crisis and integrates its climate goals into its most impactful urban decision-making processes; (4) The city and its mayor demonstrate global climate leadership and inspire others to take action supporting the Paris Agreement. C40 categorizes membership into four types: Steering Committee Cities, Megacities, Innovation Cities, and Observer Cities. Overall, C40's membership门槛 (threshold) is relatively high, with member cities generally being large in scale, having higher levels of economic development, and possessing clear emission reduction plans or actions. These standards underscore C40's commitment to building a coalition of major cities, thereby amplifying urban leadership in climate governance and strengthening identity and internal cohesion through membership.

C40 has established 7 major networks and 18 thematic networks covering areas such as energy, water, solid waste management, and transport. For knowledge and technical exchange, C40 has built an information and data-sharing platform where cities can access relevant reports and case studies to learn more effective governance approaches. C40 collates innovative practices and related initiative reports from various cities across sectors like buildings, energy, renewables, and transport, featuring them in dedicated sections on its website. It has repeatedly published climate action reports aimed at fostering discussion and broader cooperation among global megacities on solving environmental problems and creating low-carbon cities. A recent C40 report indicated that approximately one-third of the actions taken by C40 cities were realized through knowledge sharing—i.e., adopting measures already implemented in other cities.

External Operational Mechanisms

Transnational city networks typically focus on three aspects in their external operations: (1) Influencing government and intergovernmental actors; (2) Cooperating with non-governmental actors and other city networks; (3) Promoting policy implementation among member cities.

At the global level, taking C40 as an example, the city collective participates in international climate negotiations as a unique stakeholder, leveraging the influence of major cities to encourage more ambitious national climate commitments and shape the global agenda. For instance, during COP15 in Copenhagen in 2009, C40 collaborated with the host government to organize the Copenhagen Climate Summit for Mayors, attended by representatives from 80 cities, amplifying the voice of the city level in the global climate negotiation process. Through cooperation and exchange with international institutions like the UN and the World Bank, C40 can participate more influentially in global governance processes.

At the city level, C40, as a coalition of leading large cities, fosters more climate initiatives and commitments aligned with the 1.5 °C goal by convening summits, providing reports, and offering cooperative mechanisms. Through the connecting point of mayors, the city alliance can elevate urban climate ambition. It focuses specifically on sectors where city governments have the capacity to set policies and undertake transformational actions: reducing transport emissions, improving energy efficiency, while simultaneously enhancing capacities to address climate impacts.

At the non-state actor level, C40 extensively collaborates with NGOs, multinational corporations, foundations, and other entities. This not only allows it to influence these actors but also strengthens its own influence through the collaboration process. In 2011, C40 formed a partnership with the Clinton Climate Initiative (CCI), launched by former U.S. President Bill Clinton, bringing substantial resources and infrastructure. This partnership organized mayors from major cities worldwide to actively address climate change with the help of the Clinton Foundation, also establishing an organizational model for C40's collaboration with external funds.

b) **Specific Practices**

Transnational city networks are the most important link for transnational cooperation among local governments, playing a unique role in scaling up climate action. City climate networks are characterized by flexibility, pragmatism, cooperation, and efficiency. Because city cooperation does not require complex negotiations and legal procedures, the construction of transnational city climate networks and the cooperation and implementation of climate actions at the city level are considerably easier than at the national level, favoring the promotion of climate action by city alliances. Actions taken by C40 to promote climate action and expand global impact concentrate on three areas: sharing resources, influencing the agenda, and implementing action.

C40 has built a platform for exchange among cities. Based on experiences from different cities and data reports provided by C40, members analyze their own climate governance potential and feasibility, thereby formulating suitable climate policies and promoting the implementation of urban climate governance actions. C40 helps cities develop standards, processes, and tools, provides training and technical support, and assists cities in identifying optimal emission reduction methods and evaluating their effectiveness [34]. Climate initiatives launched through the C40 platform are more readily adopted and implemented by cities. For example, C40 created a city climate budget management system, which most member cities have incorporated as a priority in policy formulation and action implementation. In the seven years since world leaders signed the Paris Agreement, no single country has developed a plan consistent with the 1.5 °C target, yet 62 C40 cities are already ahead of their national governments. These C40 cities have inclusive and resilient climate action plans aligned with limiting global warming to 1.5 °C and are implementing these commitments with concrete actions. Due to efforts by cities worldwide, air pollution has decreased by 5% [35].

Taking the "Net Zero Carbon Buildings Declaration" as an example, C40 stated in this declaration that achieving the Paris Agreement's 1.5 °C goal depends critically on net-zero carbon buildings, as the building sector is a major source of GHG emissions. By the time of the declaration's launch, 23 C40 member cities had signed it, committing to ensuring that all new buildings operate at net-zero carbon by 2030 and all buildings by 2050. Signatory cities, following the standards outlined in the declaration, refer to guiding cases and data on the C40 platform, develop action roadmaps, initiate climate projects, and regularly report on project progress and carbon emission status. For instance, Cape Town's measures included promoting the adoption of stricter energy efficiency targets and the use of renewable energy in new buildings, while the city was also enhancing institutional capacity, conducting awareness campaigns, and developing financing mechanisms. Copenhagen, based on national building energy regulations for new buildings, also focused on existing buildings, decarbonizing electricity and heating through utilities, with the city's building decarbonization goal being a 40% energy saving by 2025 (compared to 2010) [36]. The participation rate for C40's five key declarations is 55.67% [37], meaning only slightly over half of the cities committed to more responsible climate action, with participation rates generally higher among cities in developed countries than in developing countries.

In January 2016, C40 launched the "Deadline 2020" programme at the C40 Mayors Summit in Mexico City. This programme aimed to support member cities in developing climate action plans that demonstrated the level of ambition and action required to achieve the Paris Agreement goals.

c) **Shortcomings**

Since its inception, C40 has been committed to providing a platform for global megacities to participate in climate governance, establishing governance norms, disseminating best practices, and facilitating experience sharing and policy learning among cities. As one of the representative city networks globally, it has achieved significant success in enhancing the participation and influence of cities as climate stakeholders. However, the following issues persist:

North–South Divide

C40 ensures member engagement and governance capacity through high entry standards. However, judged by the criteria of encompassing more cities and exerting broader influence on urban climate action, C40 evidently has shortcomings. Its organizational mechanisms also exhibit a "North–South divide," with inequalities in voice and management power. Most of C40's "leading cities" are from developed countries, and individuals from developed country cities hold a larger proportion of management positions. To date, only one of C40's seven Chairs has been from a "Global South city." Increasing participation from "Global South cities" is necessary for C40 to achieve its organizational goals and mission. When the Mayor of Rio de Janeiro served as Chair (2014–2016), cities from the Global South increased rapidly during his tenure, enabling C40 to build a broader global city network and increase its influence.

Lack of Binding Force

C40 incentivizes member city action through awards and standards but lacks constraints and supervision over the climate actions of cities after they obtain membership. This may result in reduced proactiveness and suboptimal climate outcomes among some members post-accession. C40's mechanisms include rules for exiting the network if standards are not met, but beyond this, there are no more effective measures to bind city actions. Consequently, cities failing to act are gradually marginalized or even leave the network, which also contradicts C40's aim to expand climate action.

12.6 Conclusion

As António Guterres stated, the battle for climate will be won or lost in cities. Cities will host over 70% of the world's population in the future, concentrating economic, transport, industrial, and various social activities, making them among the most critical actors in energy consumption and GHG emissions. The imperative for city engagement in climate governance lies in the fact that if cities cannot achieve net-zero emissions, the 1.5 °C goal remains an empty promise; therefore, cooperation among cities is essential.

Transnational city networks bring global cities together through a voluntary, mutually beneficial, and communicative network structure, providing a new space for urban participation in global governance and, to some extent, addressing the implementation gap in climate governance. Compared to state actors, city alliances place greater emphasis on sharing technology, experience, and knowledge, thereby avoiding various political contradictions. Unlike non-governmental actors, cities possess a degree of policy autonomy and can mobilize national resources to implement climate actions. Based on the analysis of C40's mechanisms for participating in climate governance, the internal management and external operational mechanisms of city alliances enable information and technical exchange and cooperation among cities, expand urban climate influence, and thus more actively leverage their own resource, informational, and technological advantages.

In summary, integrating cities and their networks into long-term national strategies represents a key opportunity to accelerate the implementation of climate action and achieve the goals of the Paris Agreement. For cities, actively participating in transnational climate governance networks is undoubtedly an effective strategy for enhancing their capacity to address climate vulnerability.

References

1. Decision 1/CP.21, Adoption of the Paris Agreement, FCCC/CP/2015/10/Add.1
2. Liu X, Ge Y, Hu W (2019) The geopolitical spatial influence and realization paths of non-state actors. Prog Geogr 38(11):1735–1746
3. Li X (2019) Analyzing the multidimensional influence of American non-state actors in global climate governance. Pac J 27(6):73–90
4. Zhang H (2022) The course of global climate governance and the path of sustainable development. Contemp World 6:15–20
5. Zhou Z (2016) Analysis of the Role of Multinational Corporations in Global Climate Governance. Master's thesis, Jilin University
6. Gurkov I, Filinov NB (2022) Digitalization and corporate parenting styles of multinational corporations. Int J Organ Anal 31(6):2359–2371
7. Zhang X (2021) Research on the fragmented innovation and development of global climate governance mechanisms. Popul Stand 19:3
8. Xu P (2012) Global Climate Governance Beyond Governments: A Case Study of Multinational Corporations. Master's thesis, Soochow University
9. Li L (2013) Public-Private Partnerships in Global Climate Governance. Current Affairs Press, Beijing
10. Xie F, Wang P (2023) Analysis of the role of multinational corporations in climate governance: a case study of shell. Int Public RelatS 13:14–16
11. Pang Y (2018) An Ecological Discourse Analysis of Corporate Social Responsibility Reports: A Case Study of Apple Inc. Master's thesis, Sun Yat-sen University
12. Based on Shell Quest CCS project data and public reports
13. Du Z (2022) Holding Multinational Corporations Accountable for Climate Change: A Comment on 'Friends of the Earth et al. v. Royal Dutch Shell'. Jurisprudence 8(1):393–403, 412–413
14. Du Z (2022) Holding Multinational Corporations Accountable for Climate Change: A Comment on 'Friends of the Earth et al. v. Royal Dutch Shell'. Jurisprudence 8(1):393–403, 412–413. (Note: This source appears repeated for [13] & [14] in the original text. A different source might be intended for [14], such as the BloombergNEF report itself)
15. Huang Y, Li S, Wu M, Tang X (2021) Lessons from Shell's defeat for Chinese enterprises. Int Pet Econ 29(12):69–74
16. McGann JG (1992) Academics to ideologues: A brief history of the public policy research industry. Polit Sci Polit 25(04):734–740
17. Wilcoxen PJ, McKibbin WJ (1995) Environmental Policy and International Trade. The Brookings Institution. https://www.brookings.edu/research/environmental-policy-and-international-trade/. Accessed 23 Oct 2024
18. Pan S (2019) The Involvement of the Brookings Institution in the Formulation of Climate Change Policy during the Obama Administration. Master's thesis, Beijing Foreign Studies University
19. World Resources Institute (2011) World Resources Report 2010–2011: Decision Making in a Changing Climate. World Resources Institute, Washington, DC
20. Galtung J (1964) Foreign policy opinion as a function of social position. J Peace Res 1(3):206–231
21. Ruhlman MA (2015) Who Participates in Global Governance? States, Bureaucracies, and NGOs in the United Nations. Routledge, London, pp 9–11
22. Yu H (2019) Partnership networks in global climate governance and the role of non-governmental organizations. Pacific Journal 27(11):14–25
23. Delmas MA, Young OR (2009) Governance for the Environment: New Perspectives. Cambridge University Press, Cambridge, p 7
24. Hou J, Wang Q (2013) Non-governmental organizations in international climate negotiations: status, influence, and dilemmas. J Cap Norm Univ (Soc Sci Ed) 2:55–60

25. Li X (2023) Non-Governmental Organizations and Global Climate Governance: A Perspective of Governance Embeddedness. China Social Sciences Press, Beijing
26. Li X, Wang B (2018) International non-governmental organizations and global climate governance. Glob Rev 5:136–156
27. Xu B, Ye J (2011) An analysis of the role of NGOs in addressing global environmental and climate change issues. J Shanghai Adm Inst 1:79–88
28. Zhang S (2019) A Study on the Participation of Non-Governmental Organizations in Climate Governance. China Meteorological Press, Beijing
29. Chen Y, Zhu Z, Wang G, Liu H (2022) Challenges and countermeasures for Chinese NGOs participating in international cooperation on climate change response. Environ Prot 50(7):74–77
30. IPCC (2014) AR5 Synthesis Report: Climate Change 2014. IPCC, Geneva
31. Li X, Ren X (2011) Transnational City Climate Networks in Global Climate Governance: A Case Study of C40. Journal of International Relations (6):45–51. [Note: Specific journal title from original Chinese citation is needed for accurate rendering].
32. Peng Q (2017) A Study on Transnational City Networks under Global Governance. Master's thesis, Zhejiang University, p 13
33. Ruan S (2022) Enlightenment from the Construction of Transnational City Networks in the European Region. Master's thesis, China Foreign Affairs University, p 22
34. Peng Q (2017) A Study on Transnational City Networks under Global Governance. Master's thesis, Zhejiang University, p 33
35. C40 Cities (2019) Annual Report 2019. C40 Cities Climate Leadership Group, London
36. Ruan S (2022) Enlightenment from the Construction of Transnational City Networks in the European Region. Master's thesis, China Foreign Affairs University, p 35
37. C40 Cities (2020) Annual Report 2020. C40 Cities Climate Leadership Group, London

Chapter 13
The WHO and Health: Integrating Climate and Public Health Governance

The World Health Organization (WHO) is increasingly ubiquitous in the climate governance. This chapter attaches importance to WHO's role in promoting collaborative governance of health and climate, following the logic of "discourse shaping, mechanism building, and implementation constructing". Second, it evaluates the effectiveness of WHO's engagement from mechanism and implementation level. The outcomes of this study provide some useful insights for stakeholders and policymakers in climate change and global governance, and better mobilizing the participation of governance actors in other areas on climate issues.

13.1 Introduction

Environmental issues are intimately related to human health. Since the recent industrial revolution, the globalization of environmental challenges has accelerated, and its effects on public health and human health have become more substantial. Climate change is the greatest environmental challenge facing the world today. If the linkages between environment and health in global health governance are not fully understood, it will be difficult to effectively address the increasingly complex challenges of global health governance if simply discussing health in terms of health.

In the context of global governance, there is a close interaction between the seemingly isolated mutual problems, and the paths of their respective governance will also be twined. Recognizing the significance of undertaking governance action across issues, a rising number of international organizations promote linking the organization's focus areas and other concerns in governance practice. As a multifaceted issue, climate issues have drawn international organizations to participate in discussions about development, health, immigration, security, and many other topics, which have also become a significant case study for research on the independence of international

X. Zou, *Thirty Years of Global Climate Governance*, Energy, Climate Change & Sustainability in the Global South, https://doi.org/10.1007/978-981-95-8457-4_13

organizations in emerging issues [1]. Since the 1990s, the World Health Organization has actively participated in the global climate governance process, focusing on climate and health issues.

As the most influential international organization in the field of health, the World Health Organization (WHO) has long been concerned with the linkage between climate and health issues since the 1990s. It actively participated in the global climate governance process, and the limited engagement in climate governance is an important case for the relationship between health and climate governance as well as its synergistic governance status. In light of WHO's enhanced governance leadership in the era of epidemics and other circumstances, a case study of WHO will not only help to understand the interaction between health and climate governance mechanisms, but also provide policy insights on how to promote synergy in the governance of the two issues in the post COVID-19 recovery era.

13.2 Content Analysis: Linkage Between Climate Change and Health Issues

According to the analysis above, the linkage between climate change and health issues includes two aspects, the topic content connection and the governance mechanism connection. Academics have extensively explored in-depth the connections between concerns related to health and climate change, and the global climate governance and health governance systems also reflect a consensus on issue linkage to some extent.

13.2.1 Topic Content Connection Between the Two Issues

From the problem itself, the impact of climate change on health governance has been demonstrated by numerous studies. On the one hand, climate change has a significant implication on and shapes terrestrial and oceanic ecosystems, endangering human health directly through ecological issues brought on by temperature changes, severe weather events, and ultraviolet radiation from ozone depletion, such as food and water-borne illnesses and deteriorating air quality [2]. Further analysis found that, per 10 μ g/m3 elevation of PM2.5, the risk of stroke, ischemic stroke was increased by 13%, 20%, and 12% for hemorrhagic stroke respectively. Indirectly, on the other hand, climate change affects human health through ecosystems and human society including nutrition, occupational health, mental health, and violent conflict. According to the Chinese Burden of Disease Study, air particulate matter pollution ranks fourth among the risk factors responsible for the increased disease burden, after hypertension, smoking and a high-sodium diet [3]. Moreover, a considerable number of WHO publications have pointed to climate change as a major cause in declining health worldwide. According to their research, climate change is estimated to cause approximately 250,000 fatalities per year from starvation, malaria, diarrhea, and heat stress between 2030 and 2050, with direct health damages reaching $2 billion to $4 billion per year by 2030 [4].

Based on the review of IPCC Evidence 2022, impacts to health are increasing. Climate risks are appearing faster and will become more severe sooner than previously expected, and it will be harder to adapt with increased global heating [5]. In addition, in the Six Assessment Report (AR6) of IPCC, it reinforces evidence on climate change and health; strengthens integration of social, environment and economic considerations; and emphasizes climate change equity and justice. Information from Working Group II (WG II; Adaptation) and WG III (Mitigation) takes a health perspective on the scope of the climate problem, summarizes knowledge on climate change impacts to environmental determinants of health (food, water, air) and discusses actioning this evidence in adaptation and mitigation strategies and climate change negotiations [6]. WHO Director-General's remarks at the opening ceremony of the COP27 Health Pavilion indicates that, climate change presents the health sector with a triple challenge: First, the challenge of dealing with the increasing burden of climate-fuelled disease; Second, the challenge of building climate-resilient health systems that can withstand more frequent and severe weather events; And third, the challenge of building net zero, climate-friendly health systems, while at the same time ensuring that all health facilities have a reliable and clean source of energy.

13.2.2 Governance Mechanism Connection Between the Two Issues

Currently, the effectiveness of the global climate governance mechanism is hampered by the substantially falling willingness of states to govern, and the impact of the COVID-19 pandemic has a further adverse effect on the availability of resources for climate governance. Exploring the active role of international organizations engaged in cross-issue governance in climate governance can, to a certain extent, provide new ideas for the innovation of the global climate governance mechanism, and organically integrate climate governance into the governance mechanisms of other issues, which is also conducive to mobilizing the participation of governance subjects in other issue areas on climate issues. At the level of governance mechanisms, the consensus to promote synergy between health and climate governance is gradually strengthening, and the existence of pertinent mechanisms makes it possible to realize governance synergy between the two themes. For achieving a win–win situation, it is essential to investigate synergistic actions of health and climate governance.

International organizations, often seen as the backbone of global governance, could utilize many advantages that other subjects cannot, such as the advantage of expertise, the ability to develop and disseminate norms, the capacity to implement projects across national borders, and so on. Issue connectivity is a great opportunity to expand its autonomy and influence. With the emergence of new issues and mandates from pre-existing organizations, it is generally difficult for the collective initial mandates contracted to fully cover the cross-cutting governance actions of international organizations in issue areas, and the formation of a new international organization to address the governance needs of a particular issue is typically not the best option. Although the establishment of international organizations is usually issue-specific and lacks formal mandates for cross-cutting issue areas, along with the accentuation of the need for governance actions across issue areas, more and more international organizations are going beyond their initial mandates to exert autonomous influence in other issue areas and promote the integration of their target areas with governance practices on other issues. Therefore, the analysis of autonomous actions of international organizations in new issue areas based on issue linkages has theoretical research value and therefore can contribute to better understand the role of international organizations in global governance and has theoretical research value.

In Article one of UNFCCC, it defines the adverse effects of climate change as "changes in the physical environment or biota resulting from climate change which have significant deleterious effects on the composition, resilience or productivity of natural and managed ecosystems or on the operation of socio-economic systems or on human health and welfare" [7]. The 2021 Global Health and Climate Change Conference was held during the COP26, which brought together the key actors of public health and climate change policy, committed to incorporating public health and climate justice into UN climate negotiations, and called on governments, businesses, and social institutions to promote green, healthy, and resilient outbreak recovery, highlighting the health synergies of climate policy [8]. For WHO, it is not only one of many observer organizations to the IPCC process, but also one of many organizations contributing scientific input to the production of IPCC assessments, and uses the resulting reports to inform its own work and inform guidance for national governments and the wider health community. On the other hand, the outbreak of COVID-19 has had an impact on the availability of climate governance materials, prompting more reflection on the relationship between issues and governance logic.

It can be seen that the linkages between climate and health issues are prominent at both two levels, which served as a basis for autonomous action and influence by international organizations at the intersection of climate and health issues. In the next section, the analysis will zoom into the extensive influence as well as the positive role of WHO, through its long-term commitment in promoting issue synergy.

13.3 Implementation: WHO Actions in Promoting Synergies Between Health and Climate

The World Health Organization is the specialized agency for global health governance, established by the United Nations in 1948 with the primary goal of ensuring universal access to the highest possible standard of health. As one of the most authoritative bodies in the field of global health, the WHO has also played an important role in promoting coordinated governance of climate and health issues through strategic planning and action plans over the past 60 years. The following analysis of WHO's role in the cross-cutting area of issues will be at the discourse, mechanism, and implementation levels.

13.3.1 At the Discourse Level

The human race is the sole protector of the environment, with the capability to plan wisely, to conserve providently, and to develop prudently. Conversely, it is also capable of polluting or even destroying the environment through greed, ignorance, or indifference. Given WHO's professionalism and organizational clout in global governance, its discourse promotes the international community conversation and attention on the relationship between concerns, thereby shaping the international consensus on the connection between Climate Change and Health Governance.

Gathers experts to demonstrate the relationship between climate and health issues and publish related reports.

Since the early 1990s, WHO officials have started conducting research on the relationship between climate and health risks, and published the first report on the topic with a preliminary assessment of the health impact of climate change, which laid the scientific foundation for further climate health risk research thereafter [9]. In the context, it designed to assist decision-makers and public health planners in determining the potential health problems that may arise in their region from global climatic changes caused by an increase in "greenhouse gases" and a decrease in the ozone layer. In 2007, WHO worked with Oxfam, CAR E International and International Federation of Red Cross and Red Crescent Societies, established The Global Humanitarian Fund to examine the impact of climate change on humans, and published an important report The Anatomy of a Silent Crisis before the 2009 UNFCCC summit [10]. In 2021, the WHO released a report named Climate Change and Health Research: Current Trends, Gaps and Future Outlook, which comprehensively reviewed the existing literature on climate and health, analyzing the trends and shortcomings of the current research between 2008 and 2019. This report also presents recommendations for future research, with the aim of improving the evidence base for policies to protect health from climate change [11].

Indicate the Direction of Its Organizational Governance Actions

Tedros Adhanom Ghebreyesus, the WHO Director-General, listed the impact of climate and environmental change on health on 1 July 2017 when he took office. Before the 2019 summit, he said in a statement, "climate change is one of the decisive problems of our time, is also one of the world's most urgent health threats to the country", while noting that "take measures to address and mitigate the impact of climate change, and ensure that everyone can need the time and place to get the required health services, this is the two priorities of global public health" [12]. WHO's current Director General, Mr. Tandse, has made the "health effects of climate and environmental change" one of the organization's five key priorities since July 1, 2017. It can be concluded that WHO's leading position in global health governance makes it play a crucial role in guiding the subjects of global health governance, thus successfully contributing to the evolution of health governance logic.

13.3.2 At the Mechanism Level

WHO pays much attention to the input of advancing synergistic governance at the institutional level, setting up corresponding internal bodies through numerous resolutions, strategies and work plans at the organization level, and simultaneously providing policy advice to policymakers through establishing cooperation mechanisms and launching global initiatives in interaction with other actors, in order to consolidate a global community of awareness focusing on the linkages between climate and health issues.

Frequently Published and Updated Resolutions, Strategies, Work Plans on Climate Change and Health

On the one hand, the WHO has repeatedly issued and updated the organization's resolutions, strategies and work plans on climate change and health, and has create policy provisions for how the organization is committed to achieving climate and health governance synergy.

Within WHO, multiple cross-sectoral collaborations exist between the climate change team and relevant areas of work, such as air pollution, urban health, energy, gender, nutrition and universal health coverage. The Ministry of Environment, Climate Change, and Health, on the other hand, collaborates with WHO to advance healthier environments, strengthen primary prevention, and influence public policies in all spheres. It also works to improve leadership, governance, and coordination in the health sector, create political and social support networks, and track progress toward the SDGs in order to address the underlying environmental factors that endanger health. For instance, on May 6, 2021, the WHO unveiled a climate adaptation strategy. Given that the capacity of health systems to prevent, adapt to, and address growing health hazards as well as being susceptible to shocks caused by more frequent weather events are being stressed and shocked as a result of climate change. In order to protect the health of populations from the effects of climate change and avoid widening health inequities, countries must build climate resilient health systems. Governments can join the initiative by committing to conduct climate change and health vulnerability and adaptation assessments (V&As); develop a health National Adaptation Plan (HNAP); or by allocating, or applying for, climate finance for health action. In addition to commitments under the AAC, governments are encouraged to develop an action plan or road-map for achieving a sustainable low carbon health system that minimizes climate impacts and increases resilience [13].

Established of Cooperative Mechanisms With Other Actors

First of all, WHO has been working closely with other UN agencies to incorporate health concerns into the worldwide climate change agenda. Within the UN system, in 2014, the WHO and the World Meteorological Organization jointly established the WHO/VMIO Climate and Health Office, dedicated to promoting meteorological services to international health, promote technical assistance for project implementation and research and development [14]. Moreover, the WHO represents the interests of the health sector, actively participating in UNFCCC related meetings and negotiations to support discussions on climate and health relations. In 2018, the WHO issued the Health Benefits Far Outweigh the Costs of Meeting Climate Change Goals on 24th United Nations Framework Convention, recommended to use financial incentives such as carbon pricing and energy subsidies to motivate sectors to reduce greenhouse gas and air pollutant emissions, and to encourage UNFCCC partners to eliminate existing barriers to climate-adaptive health systems [15].

Regarding interactions with other non-state players, the WHO places a high priority on the mobilization of non-state actors for the promotion of governance synergies. Currently, WHO, through its Department of Environment, Climate Change and Health (ECH), the International Union for the Conservation of Nature (IUCN), and Friends of Ecosystem-based Adaptation (FEA), is establishing an expert working group focused on biodiversity, climate, whole health, and nature-based solutions [16].

13.3.3 At the Implementation Level

In order to promote synergistic implementation of climate and health governance in project implementation, WHO is very interested in incorporating climate risk considerations in health project implementation to facilitate the assessment of climate-related risks. However, WHO's progress at the implementation level has remained largely in the publication of national voluntary assessment reports, development of action guidelines and related policy recommendations, implementation of specific projects is still relatively limited. Here selected two representative cases for analysis.

Case 1: WHO and UNDP Partner on GEF-Funded Project on Health Adaptation to Climate Change

As early as 2010, WHO actively integrated climate risk into its programs and projects, launching the first global project on public health adaptation to climate change in collaboration with the United Nations Development Programme (UNDP). The Global Environment Facility (GEF) approved US$4.5 million funding for a joint global project by the WHO and the UNDP on "piloting health adaptation to climate change [17]." The project aims to enhance early warning; build capacity; pilot specific intervention measures; and share lessons learned on protecting health from a range of hazards associated with climate change. Through the mechanism level, it brought together representatives from the seven countries where the pilot is being implemented, namely Barbados, Bhutan, China, Fiji, Jordan, Kenya and Uzbekistan. Based on the GEF-funded, the project has a bright future. Whether building on reinforced momentum and partnerships within the project or promoting the sustainability of project activities within the national system and scaling up from pilot to comprehensive programme scale, this is a meaningful innovative practice without doubt.

Case 2: Health and Climate Change Country Profiles

The WHO UNFCCC Climate and Health Country Profile Project provide country-specific estimates of current and future climate hazards and the expected burden of climate change on human health; identify opportunities for health co-benefits from climate mitigation actions; track current policy responses at national level [18]. Over 80 countries have participated in the Health and Climate Change Country Profile initiative since it began in 2015. The profiles are organized in three separate series: the current 2021–2022 global series; the 2020–2021 Small Island Developing State series; and the 2015–2018 global series.

13.4 Evaluation: The Constraints of WHO in Promoting Synergies Between Health and Climate

Since the 1990s, WHO has been actively engaged in promoting international awareness of the linkages between climate and health issues and has been a key player in promoting synergies between climate and health governance in the health sector. However, while the successful shaping of the linkages at the discourse level is an outstanding achievement of WHO's engagement in climate issues, the lack of clarity on the synergies between the two issues at the mechanism level and the lack of corresponding resources at the implementation level have limited the effectiveness of its engagement in climate issues, which is not conducive to WHO's further influence in promoting synergistic governance of the two issues.

13.4.1 Mechanism-Level Constraints

The Limited Direct Relevance of the Two Topics

While some authoritative climate reports acknowledge the impacts of climate change on human health, and the Paris Agreement was once seen as an important opportunity for WHO to access adaptation funding, the core rules of the global climate governance regime, including the United Nations Framework Convention on Climate Change, and the Kyoto Protocol, do not clearly articulate the linkages between climate action and health governance. The objective of the UNFCCC is to achieve stabilization of greenhouse gas concentrations in the atmosphere at a level that will prevent dangerous human interference with the climate system (Article 2). Despite inclusion of human health and welfare in the definition of adverse effects of climate change (Article 1), no UNFCCC decision or outcome document explicitly recognizes in its principles that stabilization and reduction of greenhouse gases will promote positive health outcomes for the planet. Without an identified reference to or role for the health sector in climate negotiations, the health stakeholder community's ability to contribute to global policy frameworks on this problem is restricted. In addition, core elements of climate governance, such as "greenhouse gas reduction" and "climate finance and technology cooperation," have limited direct relevance to health governance objectives.

Lack of Funds, Personnel and Other Supporting Facilities

The disparity between resource inputs and specific needs imposes additional restrictions on WHO's capacity-building efforts in this subject area. A report by the Green Climate Fund (GCF) analyzed the allocation of funds through five large multilateral adaptation funds in 2020, including the Adaptation Fund (AF) and the Least Developed Countries' Fund (LDCF), and found that health issues received the smallest share of funding compared to other problem areas. Reports from the NDC Partnership, NAP Global Network and from the Green Climate Fund indicate that health is one of the sectors where demands for support most frequently are left unmet. Whereas approximately two-thirds of national commitments to address climate change (NDC to the Paris Agreement), cite the importance of health and 50% of countries have strategies on climate change and health, a much lower proportion of countries have developed health adaptation plans, and less than 2% of multilateral climate finance is allocated to health projects [19]. Therefore, a significant increase in funding is needed in this regard to address the health impacts of climate change while strengthening health resilience. While the causes of climate change are global, the health impacts of climate change are local in nature. Countries with relatively low levels of economic development are often more vulnerable to health threats from climate risks due to differences in geography, infrastructure and health systems. Health systems and climate action in different countries are often confined within their borders, thus, the limited personnel and funding on climate issues within organizations restrict WHO's capacity to promote synergies between countries on climate and health governance. While WHO stated in the Project Budget From 2020 To 2021 that it would follow up the implementation and monitoring of various international

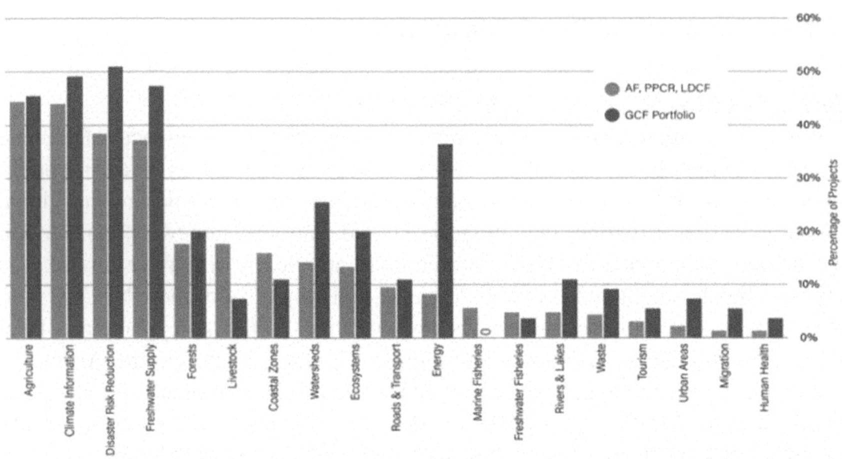

Fig. 13.1 Focus area concentration—comparing portfolios of the climate funds. (*Source* WRI analysis of AF, PPCR, and LDCF projects [232], and adaptation and cross-cutting projects in the GCF portfolio [55 as of April 2018])

agreements such as the Paris Agreement to ensure health goals and use governance mechanisms to promote healthier populations, references to climate issues are often confined to a broad representation of context and plans, without a clear road-map for specific projects (Fig. 13.1) [20].

13.4.2 4.2. Implementation-Level Constraints

From the external point of view, under the complex circumstances of the international environment, health problems are intertwined with the undercurrent of countries seeking hegemony and pursuing egotistical interests, and there is a security dilemma that impedes global health cooperation. Among them, there is considerable room to use the IPCC AR6 assessment of health evidence to inform national implementation. Fulfillment of the UNFCCC and its Paris Agreement depends on national implementation of its decisions, and the two essential elements contributing to national implementation are national adaptation plans (NAPs) and nationally determined contributions (NDCs).

In terms of NAPs, they are voluntary exercises submitted to the UNFCCC. They can, and do, include health components. The health components of NAPs can be expanded significantly to encompass and communicate health risks and actions. Explicit links between environmental changes and direct and indirect physical and mental health conditions can be included and made more comprehensive. Roadmaps can be outlined to advance interministerial agreements on cross-sectoral work and to identify preparedness plans for urgent and long- term health risks in the health sector.

In terms of NDCs, they are obligatory reports outlining countries' emissions-reduction ambitions to the UNFCCC. NDCs can include elements that strengthen the argument for ambition, and can include co-benefits such as health gains. NDCs could be strengthened with reference to whether country-level engagement on health is sufficient to address health adaptation needs. NDCs should also demonstrate co-benefits from mitigation activities with synergies to SDGs 2 and 3; from demand-side shifts in material and emissions footprints; and from other actions that increase community health, well-being and resilience, and support a low-carbon health sector.

Overall, health planning and implementation need to urgently advance beyond development of early warning systems to more comprehensive climate-health risk management at the national level. WHO-led vulnerability and adaptation assessments and ministry of health-led health national adaptation plans (HNAPs) are fundamental to multisectoral health governance, avoiding maladaptation, and incorporating equity considerations, all of which are essential to building resilience [21].

It can be seen that the further development of WHO's role in promoting synergies between climate and health governance needs to rely on the construction of supporting mechanisms and the guarantee of a stable supply of resources. On the one hand, WHO needs to further maintain the visibility of health issues in the global climate negotiations and promote the inclusion of health governance objectives in NDCs. At the same time, WHO should concentrate on forming partnerships with climate finance-related mechanisms and promoting the importance and investment of relevant financial mechanisms in health governance, so as to strengthen the organization's position at the intersection of climate and health, and play a greater role in climate governance through staffing and financial security.

13.5 Conclusion and Discussion

In the context of globalization, the close connection between issues under the epidemic has become more prominent, and the realization of synergistic governance is of great significance to improve the effectiveness of global governance. With the further improvement of global climate governance mechanisms and health governance structures after the epidemic, achieving synergy between climate and health governance will become the key issue in the global governance agenda with broad participation of multiple governance actors, in which the leadership of international organizations in shaping inclusive, legitimate and fair governance mechanisms will be further highlighted. WHO, as the most influential global public health authority in the international community, plays a leadership role in global health with its health expertise, sets the global health agenda and norms, and facilitates the achievement of international cooperation in health. At the same time, WHO is crucial in advocating coordinated government action on pressing health and climate concerns. In the area of health governance, WHO has long been involved in forming the connections between climate and health challenges and has made strides toward promoting synergistic governance of the two issues.

However, the current lack of global governance in non-traditional areas such as global health and climate is still prominent, and the supply of public goods in cross-cutting areas is even more scarce. At the discourse level, WHO should make full use of its discursive influence to convene relevant experts to make scientific arguments on economic recovery and climate governance issues, further maintain the visibility of health issues in global climate negotiations, and promote countries to give high priority to health governance goals into the NDCs. On the other hand, implementation-level accomplishments are modest primarily due to the absence of pertinent regulations, the power of different countries, and the varying stages of the epidemic. WHO should play an active role in the international negotiations and guide the parties to fully consider the governance objectives of these two issues when formulating the relevant rules. In global issues, different countries suffer from different degrees of negative impacts and have different capacities to take specific actions. For countries with weaker governance capacity, WHO is ought to mobilize other countries to provide corresponding technical and financial support to enable them participate in global climate governance, enhance the risk-resistance of their national economic systems, and ultimately promote the achievement of global governance goals. The outbreak of the COVID-19 has brought more attention to WHO's role in global risk response, and the post-epidemic recovery phase is expected to be the crucial opportunity for WHO to further participate in global governance and promote synergy between climate and health governance.

References

1. Watts N, Amann M, Arnell N, et al (2019) The 2019 report of The Lancet Countdown on health and climate change: ensuring that the health of a child born today is not defined by a changing climate. Lancet 394(10211):1836–1878. https://doi.org/10.1016/S0140-6736(19)32596-6. Accessed 12 Dec 2024
2. Ricciardi W, Marcheggiani S, Puccinelli C et al (2019) Health and climate change: science calls for global action. Ann Ist Super Sanità 55(4):324
3. 殷鹏,齐金蕾,刘韫宁．2005~2017年中国疾病负担研究报告．中国循环杂志，2019，34: 1145–1154
4. WHO (2019) Programme budget 2020–2021, p 55, 69. https://www.who.int/publications/i/item/programme-budget-2020-2021, Login time: 2022.10.14. Accessed 14 Dec 2024
5. WHO (2022) Climate change. https://www.who.int/health-topics/climate-change#tab=tab_1, Login time: 2022.10.14
6. WHO (2022) Review of IPCC evidence 2022: climate change, health, and well-being. Advanced version. WHO Team (Climate Change and Health), p 16, Advanced Version, October 2022. CC BY-NC-SA 3.0 IGO
7. IPCC AR6 WG II Full Report, Chapter 7 (WG II Ch. 7) Cissé G, McLeman R, Adams H, et al (2022) Health, wellbeing, and the changing structure of communities. In: Climate change 2022: impacts, adaptation and vulnerability. Contribution of Working Group II to the Sixth Assessment Report of the Intergovernmental Panel on Climate Change. https://www.ipcc.ch/report/ar6/wg2/downloads/report/IPCC_AR6_WGII_FullReport.pdf. Accessed 1 Jan 2023
8. UNFCCC (1992) United Nations Framework Convention on Climate Change, Article 1. https://unfccc.int/files/essential_background/background_publications_htmlpdf/application/pdf/conveng.pdf. Accessed 21 Oct 2024

9. WHO (2021) 2021 global conference on health and climate change. https://www.who.int/news-room/events/detail/2021/106/default-calendar/2021-global-conference-on-health-and-climate-change. Accessed 20 Nov 2024
10. WHO (1990) Potential health effects of climatic change: report of a WHO task group. World Health Organization, Geneva. https://apps.who.int/iris/handle/10665/62016. Accessed 22 Oct 2024
11. Global Humanitarian Forum (2009) The anatomy of a silent crisis. Global Humanitarian Forum, Geneva. http://www.ghf-ge.org/human-impact-report.pdf. Accessed 25 Nov 2024
12. Research Subcommittee of the WHO-Civil Society Working Group to Advance Action on Climate Change and Health, WHO and the Global Climate and Health Alliance (GCHA), Climate Change and Health Research: Current Trends, Gaps and Perspectives for the Future, Geneva: World Health Organization, 2021
13. WHO (2019) WHO Director-General urges world leaders to protect health from climate change. https://www.who.int/news/item/11-092019-who-director-general-urges-world-leaders-to-protect-health-from-climate-change, Login time: 2023.1.4. Accessed 24 Nov 2024
14. WHO (2021) Adaptation action coalition health launch: building climate resilient health systems. https://www.who.int/news-room/events/detail/2021/05/06/default-calendar/adaptation-action-coalition-health-launch, Login time: 2023.1.4. Accessed 14 Jul 2024
15. WMO (2023) Public health. https://public.wmo.int/en/our-mandate/focus-areas/public-health, Login time: 2023.1.4. Accessed 25 Jul 2024
16. WHO (2018) Health benefits far outweigh the costs of meeting climate change goals. https://www.who.int/news/item/05-12-2018-health-benefits-far-outweigh-the-costs-of-meeting-climate-change-goals, Login time: 2023.1.4. Accessed 29 Jul 2024
17. WHO (2021) New WHO-IUCN expert working group on biodiversity, climate, one health and nature-based solutions. https://www.who.int/news/item/30-03-2021-who-iucn-expert-working-group-biodiversity. Accessed 7 Oct 2024
18. WHO and UNDP Partner on GEF-funded Project on Health Adaptation to Climate Change | News | SDG Knowledge Hub | IISD
19. WHO https://www.who.int/teams/environment-climate-change-and-health/climate-change-and-health/evidence-monitoring/country-profiles, Login time: 2023.1.4. Accessed 10 Aug 2024
20. WHO https://cdn.who.int/media/docs/default-source/climate-change/who-initiative-on-climate-resilient-health-systems.pdf?sfvrsn=73bdfd15_5, Login time: 2023.1.4. Accessed 23 Nov 2024
21. WHO (2021) Quality criteria for health national adaptation plans. World Health Organization, Geneva

Appendices

This appendix presents excerpts from three of the most significant international climate treaties, selected from the **United Nations Treaty Collection** (https://treaties.un.org/Pages/). The purpose of this section is to provide readers with a concise yet authoritative reference to the legal foundations of global climate governance. By offering direct access to the essential texts and their current status, it aims to facilitate a clearer understanding of the evolution, structure, and operative mechanisms of these cornerstone agreements, thereby serving as both a contextual supplement and a practical reference tool for the main body of this book.

Appendix 1: The United Nations Framework Convention on Climate Change

Appendix 2: Kyoto Protocol

Appendix 3: Paris Agreement

X. Zou, *Thirty Years of Global Climate Governance*, Energy, Climate Change & Sustainability in the Global South, https://doi.org/10.1007/978-981-95-8457-4

Appendix 1

United Nations Framework Convention on Climate Change

New York, 9 May 1992

ENTRY INTO FORCE:	21 March 1994, in accordance with article 23(1)
REGISTRATION:	21 March 1994, No. 30822
STATUS:	Signatories: 165. Parties: 198[1]
TEXT:	United Nations, *Treaty Series*, vol. 1771, p. 107; and depositary notifications C.N.148.1993.TREATIES-4 of 12 July 1993 (procès-verbal of rectification of the original texts of the Convention); C.N.436.1993.TREATIES-12 of 15 December 1993 (corrigendum to C.N.148.1993.TREATIES-4 of 12 July 1993); C.N.247.1993.TREATIES-6 of 24 November 1993 (procès-verbal of rectification of the authentic French text); C.N.462.1993.TREATIES-13 of 30 December 1993 (corrigendum to C.N.247.1993.TREATIES-6 of 24 November 1993); C.N.544.1997.TREATIES-6 of 13 February 1997 (amendment to the list in annex I to the Convention); and C.N.1478.2001.TREATIES-2 of 28 December 2001 (amendment to the list in annex II to the Convention); C.N.237.2010.TREATIES-2 of 26 April 2010 (adoption of amendment to the list in the Annex I to the Convention); C.N.355.2012.TREATIES-XXVII.7 of 9 July 2012 (adoption of amendment to Annex I to the Convention) and C.N.81.2013.TREATIES-XXVII.7 of 14 January 2013 (entry into force of amendment to Annex I to the Convention)

Note: The Convention was agreed upon and adopted by the Intergovernmental Negotiating Committee for a Framework Convention on Climate Change, during its Fifth session, second part, held at New York from 30 April to 9 May 1992. In accordance with its article 20, the Convention was open for signature by States Members of the United Nations or of any of its specialized agencies or that are

X. Zou, *Thirty Years of Global Climate Governance*, Energy, Climate Change & Sustainability in the Global South, https://doi.org/10.1007/978-981-95-8457-4

Parties to the Statute of the International Court of Justice and by regional economic integration organizations, at Rio de Janeiro during the United Nations Conference on Environment and Development, from 4 to 14 June 1992, and remained thereafter open at the United Nations Headquarters in New York until 19 June 1993.

Participant	*Signature*		*Approval(AA), Acceptance(A), Accession(a), Succession(d), Ratification*	
Afghanistan	12 Jun	1992	19 Sep	2002
Albania			3 Oct	1994 a
Algeria	13 Jun	1992	9 Jun	1993
Andorra			2 Mar	2011 a
Angola	14 Jun	1992	17 May	2000
Antigua and Barbuda	4 Jun	1992	2 Feb	1993
Argentina	12 Jun	1992	11 Mar	1994
Armenia	13 Jun	1992	14 May	1993 A
Australia	4 Jun	1992	30 Dec	1992
Austria	8 Jun	1992	28 Feb	1994
Azerbaijan	12 Jun	1992	16 May	1995
Bahamas (The)	12 Jun	1992	29 Mar	1994
Bahrain	8 Jun	1992	28 Dec	1994
Bangladesh	9 Jun	1992	15 Apr	1994
Barbados	12 Jun	1992	23 Mar	1994
Belarus	11 Jun	1992	11 May	2000 AA
Belgium	4 Jun	1992	16 Jan	1996
Belize	13 Jun	1992	31 Oct	1994
Benin	13 Jun	1992	30 Jun	1994
Bhutan	11 Jun	1992	25 Aug	1995
Bolivia (Plurinational State of)	10 Jun	1992	3 Oct	1994
Bosnia and Herzegovina			7 Sep	2000 a
Botswana	12 Jun	1992	27 Jan	1994
Brazil	4 Jun	1992	28 Feb	1994
Brunei Darussalam			7 Aug	2007 a
Bulgaria	5 Jun	1992	12 May	1995
Burkina Faso	12 Jun	1992	2 Sep	1993
Burundi	11 Jun	1992	6 Jan	1997
Cabo Verde	12 Jun	1992	29 Mar	1995
Cambodia			18 Dec	1995 a
Cameroon	14 Jun	1992	19 Oct	1994
Canada	12 Jun	1992	4 Dec	1992

Participant	*Signature*		*Approval(AA), Acceptance(A), Accession(a), Succession(d), Ratification*	
Central African Republic	13 Jun	1992	10 Mar	1995
Chad	12 Jun	1992	7 Jun	1994
Chile	13 Jun	1992	22 Dec	1994
China[1,2]	11 Jun	1992	5 Jan	1993
Colombia	13 Jun	1992	22 Mar	1995
Comoros	11 Jun	1992	31 Oct	1994
Congo	12 Jun	1992	14 Oct	1996
Cook Islands	12 Jun	1992	20 Apr	1993
Costa Rica	13 Jun	1992	26 Aug	1994
Côte d'Ivoire	10 Jun	1992	29 Nov	1994
Croatia	11 Jun	1992	8 Apr	1996 A
Cuba	13 Jun	1992	5 Jan	1994
Cyprus	12 Jun	1992	15 Oct	1997
Czech Republic	18 Jun	1993	7 Oct	1993 AA
Democratic People's Republic of Korea	11 Jun	1992	5 Dec	1994 AA
Democratic Republic of the Congo	11 Jun	1992	9 Jan	1995
Denmark	9 Jun	1992	21 Dec	1993
Djibouti	12 Jun	1992	27 Aug	1995
Dominica			21 Jun	1993 a
Dominican Republic	12 Jun	1992	7 Oct	1998
Ecuador	9 Jun	1992	23 Feb	1993
Egypt	9 Jun	1992	5 Dec	1994
El Salvador	13 Jun	1992	4 Dec	1995
Equatorial Guinea			16 Aug	2000 a
Eritrea			24 Apr	1995 a
Estonia	12 Jun	1992	27 Jul	1994
Eswatini	12 Jun	1992	7 Oct	1996
Ethiopia	10 Jun	1992	5 Apr	1994
European Union	13 Jun	1992	21 Dec	1993 AA
Fiji	9 Oct	1992	25 Feb	1993
Finland	4 Jun	1992	3 May	1994 A
France	13 Jun	1992	25 Mar	1994
Gabon	12 Jun	1992	21 Jan	1998
Gambia	12 Jun	1992	10 Jun	1994

Participant	*Signature*		*Approval(AA), Acceptance(A), Accession(a), Succession(d), Ratification*	
Georgia			29 Jul	1994 a
Germany	12 Jun	1992	9 Dec	1993
Ghana	12 Jun	1992	6 Sep	1995
Greece	12 Jun	1992	4 Aug	1994
Grenada	3 Dec	1992	11 Aug	1994
Guatemala	13 Jun	1992	15 Dec	1995
Guinea	12 Jun	1992	7 May	1993
Guinea Bissau	12 Jun	1992	27 Oct	1995
Guyana	13 Jun	1992	29 Aug	1994
Haiti	13 Jun	1992	25 Sep	1996
Holy See			6 Jul	2022 a
Honduras	13 Jun	1992	19 Oct	1995
Hungary	13 Jun	1992	24 Feb	1994
Iceland	4 Jun	1992	16 Jun	1993
India	10 Jun	1992	1 Nov	1993
Indonesia	5 Jun	1992	23 Aug	1994
Iran (Islamic Republic of)	14 Jun	1992	18 Jul	1996
Iraq			28 Jul	2009 a
Ireland	13 Jun	1992	20 Apr	1994
Israel	4 Jun	1992	4 Jun	1996
Italy	5 Jun	1992	15 Apr	1994
Jamaica	12 Jun	1992	6 Jan	1995
Japan	13 Jun	1992	28 May	1993 A
Jordan	11 Jun	1992	12 Nov	1993
Kazakhstan	8 Jun	1992	17 May	1995
Kenya	12 Jun	1992	30 Aug	1994
Kiribati	13 Jun	1992	7 Feb	1995
Kuwait			28 Dec	1994 a
Kyrgyzstan			25 May	2000 a
Lao People's Democratic Republic			4 Jan	1995 a
Latvia	11 Jun	1992	23 Mar	1995
Lebanon	12 Jun	1992	15 Dec	1994
Lesotho	11 Jun	1992	7 Feb	1995
Liberia	12 Jun	1992	5 Nov	2002
Libya	29 Jun	1992	14 Jun	1999

Participant	*Signature*		*Approval(AA), Acceptance(A), Accession(a), Succession(d), Ratification*	
Liechtenstein	4 Jun	1992	22 Jun	1994
Lithuania	11 Jun	1992	24 Mar	1995
Luxembourg	9 Jun	1992	9 May	1994
Madagascar	10 Jun	1992	2 Jun	1999
Malawi	10 Jun	1992	21 Apr	1994
Malaysia	9 Jun	1993	13 Jul	1994
Maldives	12 Jun	1992	9 Nov	1992
Mali	30 Sep	1992	28 Dec lePara>	1994
Malta	12 Jun	1992	17 Mar	1994
Marshall Islands	12 Jun	1992	8 Oct	1992
Mauritania	12 Jun	1992	20 Jan	1994
Mauritius	10 Jun	1992	4 Sep	1992
Mexico	13 Jun	1992	11 Mar	1993
Micronesia (Federated States of)	12 Jun	1992	18 Nov	1993
Monaco	11 Jun	1992	20 Nov	1992
Mongolia	12 Jun	1992	30 Sep	1993
Montenegro[4]			23 Oct	2006 d
Morocco	13 Jun	1992	28 Dec	1995
Mozambique	12 Jun	1992	25 Aug	1995
Myanmar	11 Jun	1992	25 Nov	1994
Namibia	12 Jun	1992	16 May	1995
Nauru	8 Jun	1992	11 Nov	1993
Nepal	12 Jun	1992	2 May	1994
Netherlands (Kingdom of the)[5]	4 Jun	1992	20 Dec	1993 A
New Zealand[6]	4 Jun	1992	16 Sep	1993
Nicaragua	13 Jun	1992	31 Oct	1995
Niger	11 Jun	1992	25 Jul	1995
Nigeria	13 Jun	1992	29 Aug	1994
Niue			28 Feb	1996 a
North Macedonia			28 Jan	1998 a
Norway	4 Jun	1992	9 Jul	1993
Oman	11 Jun	1992	8 Feb	1995
Pakistan	13 Jun	1992	1 Jun	1994
Palau			10 Dec	1999 a

Participant	*Signature*		*Approval(AA), Acceptance(A), Accession(a), Succession(d), Ratification*	
Panama	18 Mar	1993	23 May	1995
Papua New Guinea	13 Jun	1992	16 Mar	1993
Paraguay	12 Jun	1992	24 Feb	1994
Peru	12 Jun	1992	7 Jun	1993
Philippines	12 Jun	1992	2 Aug	1994
Poland	5 Jun	1992	28 Jul	1994
Portugal (See endnote 3)	13 Jun	1992	21 Dec	1993
Qatar			18 Apr	1996 a
Republic of Korea	13 Jun	1992	14 Dec	1993
Republic of Moldova	12 Jun	1992	9 Jun	1995
Romania	5 Jun	1992	8 Jun	1994
Russian Federation	13 Jun	1992	28 Dec	1994
Rwanda	10 Jun	1992	18 Aug	1998
Samoa	12 Jun	1992	29 Nov	1994
San Marino	10 Jun	1992	28 Oct	1994
Sao Tome and Principe	12 Jun	1992	29 Sep	1999
Saudi Arabia			28 Dec	1994 a
Senegal	13 Jun	1992	17 Oct	1994
Serbia			12 Mar	2001 a
Seychelles	10 Jun	1992	22 Sep	1992
Sierra Leone	11 Feb	1993	22 Jun	1995
Singapore	13 Jun	1992	29 May	1997
Slovakia	19 May	1993	25 Aug	1994 AA
Slovenia	13 Jun	1992	1 Dec	1995
Solomon Islands	13 Jun	1992	28 Dec	1994
Somalia			11 Sep	2009 a
South Africa	15 Jun	1993	29 Aug	1997
South Sudan			17 Feb	2014 a
Spain	13 Jun	1992	21 Dec	1993
Sri Lanka	10 Jun	1992	23 Nov	1993
St. Kitts and Nevis	12 Jun	1992	7 Jan	1993
St. Lucia	14 Jun	1993	14 Jun	1993
St. Vincent and the Grenadines			2 Dec	1996 a
State of Palestine			18 Dec	2015 a
Sudan	9 Jun	1992	19 Nov	1993
Suriname	13 Jun	1992	14 Oct	1997

Participant	*Signature*		*Approval(AA), Acceptance(A), Accession(a), Succession(d), Ratification*	
Sweden	8 Jun	1992	23 Jun	1993
Switzerland	12 Jun	1992	10 Dec	1993
Syrian Arab Republic			4 Jan	1996 a
Tajikistan			7 Jan	1998 a
Thailand	12 Jun	1992	28 Dec	1994
Timor Leste			10 Oct	2006 a
Togo	12 Jun	1992	8 Mar	1995 A
Tonga			20 Jul	1998 a
Trinidad and Tobago	11 Jun	1992	24 Jun	1994
Tunisia	13 Jun	1992	15 Jul	1993
Türkiye			24 Feb	2004 a
Turkmenistan			5 Jun	1995 a
Tuvalu	8 Jun	1992	26 Oct	1993
Uganda	13 Jun	1992	8 Sep	1993
Ukraine	11 Jun	1992	13 May	1997
United Arab Emirates			29 Dec	1995 a
United Kingdom of Great Britain and Northern Ireland,[78]	12 Jun	1992	8 Dec	1993
United Republic of Tanzania	12 Jun	1992	17 Apr	1996
United States of America	12 Jun	1992	15 Oct	1992
Uruguay	4 Jun	1992	18 Aug	1994
Uzbekistan			20 Jun	1993 a
Vanuatu	9 Jun	1992	25 Mar	1993
Venezuela (Bolivarian Republic of)	12 Jun	1992	28 Dec	1994
Viet Nam	11 Jun	1992	16 Nov	1994
Yemen	12 Jun	1992	21 Feb	1996
Zambia	11 Jun	1992	28 May	1993
Zimbabwe	12 Jun	1992	3 Nov	1992

Declarations

(Unless otherwise indicated, the declarations were made upon ratification, accession, acceptance, approval or succession.)

Bulgaria

"The Republic of Bulgaria declares that in accordance with article 4, paragraph 6, and with respect to paragraph 2(b) of the said article, it accepts as a basis of the anthropogenic emissions in Bulgaria of carbon dioxide and other greenhouse gases not controlled by the Montreal Protocol, the 1988 levels of the said emissions in the country and not their 1990 levels, keeping records of and comparing the emission rates during the subsequent years."

Croatia

"The Republic of Croatia declares that it intends to be bound by the provisions of the Annex 1, as a country undergoing the process of transition to a market economy."

Cuba

With reference to article 14 of the United Nations Framework Convention on Climate Change, the Government of the Republic of Cuba declares that, insofar as concerns the Republic of Cuba, any dispute that may arise between the Parties concerning the interpretation or application of the Convention shall be settled through negotiation through the diplomatic channel.

European Union

"The European Economic Community and its Member States declare, for the purposes of clarity, that the inclusion of the European Community as well as its Member States in the lists in the Annexes to the Convention is without prejudice to the division of competence and responsibilities between the Community and its Member States, which is to be declared in accordance with article 21(3) of the Convention."

"The European Economic Community and its Member States declare that the commitment to limit anthropogenic CO_2 emissions set out in article 4(2) of the Convention will be fulfilled in the Community as a whole through action by the Community and its Member States, within the respective competence of each.

In this perspective, the Community and its Member States reaffirm the objectives set out in the Council conclusions of 29 October 1990, and in particular the objective of stabilization of CO_2 emission by 2000 and 1990 level in the Community as a whole.

The European Economic Community and its Member States are elaborating a coherent strategy in order to attain this objective."

Fiji

"The Government of Fiji declares its understanding that signature of the Convention shall, in no way, constitute a renunciation of any rights under international law concerning state responsibility for the adverse effects of climate change, and that no provisions in the Convention can be interpreted as derogating from the principles of general international law."

Holy See

"By acceding to the United Nations Framework Convention on Climate Change in the name and on behalf of Vatican City State, the Holy See intends to contribute to the efforts of all States to work together in solidarity, in accordance with their common but differentiated responsibilities and respective capabilities, in an effective response to the challenges posed by climate change to humankind and to our common home.

In light of the territorial nature of the obligations set forth in the United Nations Framework Convention on Climate Change, the Holy See declares, for the avoidance of doubt, that in acceding to the Convention only in the name and on behalf of Vatican City State it commits itself to apply its provisions exclusively within the Territory of the Vatican City State, as circumscribed by the Leonine Walls.

The Holy See, in conformity with its particular mission, reiterates, on behalf of Vatican City State, its position regarding the term 'gender'. The Holy See underlines that any reference to 'gender' and related terms in any document that has been or that will be adopted by the Conference of State Parties or by its subsidiary bodies is to be understood as grounded on the biological sexual identity that is male and female.

The Holy See upholds and promotes a holistic and integrated approach that is firmly centered on the human dignity and integral development of every person."

Hungary

"The Government of the Republic of Hungary attributes great significance to the United Nations Framework Convention on Climate Change and it reiterates its position in accordance with the provisions of article 4.6 of the Convention on certain degree of flexibility that the average level of anthropogenic carbon-dioxide emissions for the period of 1985–1987 will be considered as reference level in context of the commitments under article 4.2 of the Convention. This understanding is closely related to the 'process of transition' as it is given in article 4.6 of the Convention. The Government of the Republic of Hungary declares that it will do all efforts to contribute to the objective of the Convention."

Kiribati

"The Government of the Republic of Kiribati declares its understanding that signature and /or ratification of the Convention shall in no way constitute a renunciation of any rights under international law concerning state responsibility for the adverse effects of climate change, and that no provisions in the Convention can be interpreted as derogating from the principles of general international law."

Monaco

In accordance with sub-paragraph g of article 4.2 of the Convention, the Principality of Monaco declares that it intends to be bound by the provisions of sub-paragraphs a and b of said article.

Nauru

"The Government of Nauru declares its understanding that signature of the Convention shall in no way constitute a renunciation of any rights under international law concerning state responsibility for the adverse effects of climate change, and that no provisions in the Convention can be interpreted as derogating from the principles of general international law."

Netherlands (Kingdom of The)

"The Kingdom of the Netherlands declares, in accordance with paragraph 2 of Article 14 of the United Nations Framework Convention on Climate Change, that it accepts both means of dispute settlement referred to in that paragraph as compulsory in relation to any Party accepting one or both means of dispute settlement."

Papua New Guinea

"The Government of the Independent State of Papua New Guinea declares its understanding that ratification of the Convention shall in no way constitute a renunciation of any rights under International Law concerning State responsibility for the adverse effects of Climate Change as derogating from the principles of general International Law."

Solomon Islands

"In pursuance of article 14(2) of the said Convention [the Government of the Solomon Islands] shall recognise as compulsory, arbitration, in accordance with procedures to be adopted by the Conference of the Parties as soon as practicable, in an annex on arbitration."

Tuvalu

"The Government of Tuvalu declares its understanding that signature of the Convention shall in no way constitute a renunciation of any rights under international law concerning state responsibility for the adverse effects of climate change, and that no provisions in the Convention can be interpreted as derogating from the principles of general international law."

Notifications Made Under Article 4 (2) (g)[9]

Participant	*Date of receipt of the notification:*
Czech Republic	27 Nov 1995
Kazakhstan	23 Mar 2000
Monaco	20 Nov 1992
Slovakia	23 Feb 1996
Slovenia	9 Jun 1998

Notes

1. For the purpose of entry into force of the [Convention/Protocol], any instrument of ratification, acceptance, approval or accession deposited by a regional economic integration organization shall not be counted as additional to those deposited by member States of that Organization.

2. By a communication received on 8 April 2003, the Government of the Government of the People's Republic of China notified the Secretary-General of the following:

   "In accordance with the provisions of Article 153 of the Basic Law of the Hong Kong Special Administrative Region of the People's Republic of China of 1990, the Government of the People's Republic of China decides that the United Nations Framework Convention on Climate Change and the Kyoto Protocol to the United Nations Framework Convention on Climate Change shall apply to the Hong Kong Special Administrative Region of the People's Republic of China.

   The United Nations Framework Convention on Climate Change continues to be implemented in the Macao Special Administrative Region of the People's Republic of China. The Kyoto Protocol to the United Nations Framework Convention on Climate Change shall not apply to the Macao Special Administrative Region of the People's Republic of China until the Government of China notifies otherwise."
3. On 28 June 1999, the Government of Portugal informed the Secretary-General the the Convention would also apply to Macao.

   Subsequently, the Secretary-General received communications concerning the status of Macao from Portugal and China (see note 1 under "Portugal" and note 3 under "China" in the "Historical Information" section in the front matter of this volume.) Upon resuming the exercise of sovereignty over Macao, China notified the Secretary-General that the Convention will also apply to the Macao Special Administrative Region.
4. See note 1 under "Montenegro" in the "Historical Information" section in the front matter of this volume.
5. For the Kingdom in Europe.
6. Upon ratification, New Zealand had notified the Secretary-General of a territorial exclusion with respect to Tokealau. On 13 November 2017, New Zealand notified that it extends the application of the Convention to Tokelau. See C.N.704.2017.TREATIES-XXVII.7 of 13 November 2017.
7. In respect of Great Britain and Northern Ireland, the Bailiwick of Jersey and the Isle of Man. On 4 April 2006: in respect of the Bailiwick of Guernsey. On 2 January 2007: in respect of Gibraltar. On 7 March 2007: in respect of Bermuda, Cayman Islands, Falkland Islands (Malvinas).
8. By a communication received on 27 March 2007, the Government of Argentina notified the Secretary-General of the following:

   The Argentine Republic objects to the extension of the territorial application to the United Nations Framework Convention on Climate Change of 9 May 1992 with respect to the Malvinas Islands, which was notified by the United Kingdom of Great Britain and Northern Ireland to the Depositary of the Convention on 7 March 2007.

   The Argentine Republic reaffirms its sovereignty over the Malvinas Islands, the South Georgia and South Sandwich Islands and the surrounding maritime spaces, which are an integral part of its national territory, and recalls that the General Assembly of the United Nations adopted resolutions 2065 (XX), 3160

(XXVIII), 31/49, 37/9, 38/12, 39/6, 40/21, 41/40, 42/19 and 43/25, which recognize the existence of a dispute over sovereignty and request the Governments of the Argentine Republic and the United Kingdom of Great Britain and Northern Ireland to initiate negotiations with a view to finding the means to resolve peacefully and definitively the pending problems between both countries, including all aspects on the future of the Malvinas Islands, in accordance with the Charter of the United Nations.

9. States having, in accordance with article 4 (2)(g), notified the Secretary-General of their intention to be bound by article 4 (2)(a) and (b) of the Convention.

Appendix 2

Kyoto Protocol to the United Nations Framework Convention on Climate Change

Kyoto, 11 December 1997

ENTRY INTO FORCE:	16 February 2005, in accordance with article 25(1) and article 25(3) which read as follows: "1. This Protocol shall enter into force on the ninetieth day after the date on which not less than 55 Parties to the Convention, incorporating Parties included in Annex I which accounted in total for at least 55 per cent of the total carbon dioxide emissions for 1990 of the Parties included in Annex I, have deposited their instruments of ratification, acceptance, approval or accession." "3. For each State or regional economic integration organization that ratifies, accepts or approves this Protocol or accedes thereto after the conditions set out in paragraph 1 above for entry into force have been fulfilled, this Protocol shall enter into force on the ninetieth day following the date of deposit of its instrument of ratification acceptance, approval or accession"
REGISTRATION:	16 February 2005, No. 30822
STATUS:	Signatories: 83. Parties: 192[1]
TEXT:	United Nations, *Treaty Series*, vol. 2303, p. 162; depositary notifications C.N.101.2004.TREATIES-1 of 11 February 2004 [Proposed corrections to the original texts of the Protocol (Arabic and French versions)] and C.N.439.2004.TREATIES-4 of 12 May 2004 [Corrections to the original texts of the Protocol (Arabic and French versions)]; C.N.380.2007.TREATIES-5 of 17 April 2007 (Adoption of an amendment to Annex B of the Protocol)

Note: The Protocol was adopted at the third session of the Conference of the Parties to the 1992 United Nations Framework Convention on Climate Change ("the Convention"), held at Kyoto (Japan) from 1 to 11 December 1997. The Protocol shall be open for signature by States and regional economic integration organizations

X. Zou, *Thirty Years of Global Climate Governance*, Energy, Climate Change & Sustainability in the Global South, https://doi.org/10.1007/978-981-95-8457-4

which are Parties to the Convention at United Nations Headquarters in New York from 16 March 1998 to 15 March 1999 in accordance with its article 24 (1).

Participant	*Signature*		*Approval(AA), Acceptance(A), Accession(a), Succession(d), Ratification*	
Afghanistan			25 Mar	2013 a
Albania			1 Apr	2005 a
Algeria			16 Feb	2005 a
Angola			8 May	2007 a
Antigua and Barbuda	16 Mar	1998	3 Nov	1998
Argentina	16 Mar	1998	28 Sep	2001
Armenia			25 Apr	2003 a
Australia	29 Apr	1998	12 Dec	2007
Austria	29 Apr	1998	31 May	2002
Azerbaijan			28 Sep	2000 a
Bahamas (The)			9 Apr	1999 a
Bahrain			31 Jan	2006 a
Bangladesh			22 Oct	2001 a
Barbados			7 Aug	2000 a
Belarus			26 Aug	2005 a
Belgium	29 Apr	1998	31 May	2002
Belize			26 Sep	2003 a
Benin			25 Feb	2002 a
Bhutan			26 Aug	2002 a
Bolivia (Plurinational State of)	9 Jul	1998	30 Nov	1999
Bosnia and Herzegovina			16 Apr	2007 a
Botswana			8 Aug	2003 a
Brazil	29 Apr	1998	23 Aug	2002
Brunei Darussalam			20 Aug	2009 a
Bulgaria	18 Sep	1998	15 Aug	2002
Burkina Faso			31 Mar	2005 a
Burundi			18 Oct	2001 a
Cabo Verde			10 Feb	2006 a
Cambodia			22 Aug	2002 a
Cameroon			28 Aug	2002 a

Participant	*Signature*		*Approval(AA), Acceptance(A), Accession(a), Succession(d), Ratification*	
Canada[2]	[29 Apr	1998]	[17 Dec	2002]
Central African Republic			18 Mar	2008 a
Chad			18 Aug	2009 a
Chile	17 Jun	1998	26 Aug	2002
China[3]	29 May	1998	30 Aug	2002 AA
Colombia			30 Nov	2001 a
Comoros			10 Apr	2008 a
Congo			12 Feb	2007 a
Cook Islands	16 Sep	1998	27 Aug	2001
Costa Rica	27 Apr	1998	9 Aug	2002
Côte d'Ivoire			23 Apr	2007 a
Croatia	11 Mar	1999	30 May	2007
Cuba	15 Mar	1999	30 Apr	2002
Cyprus			16 Jul	1999 a
Czech Republic	Nov	1998	15 Nov	2001 AA
Democratic People's Republic of Korea			27 Apr	2005 a
Democratic Republic of the Congo			23 Mar	2005 a
Denmark[4]	29 Apr	1998	31 May	2002
Djibouti			12 Mar	2002 a
Dominica			25 Jan	2005 a
Dominican Republic			12 Feb	2002 a
Ecuador	15 Jan	1999	13 Jan	2000
Egypt	15 Mar	1999	12 Jan	2005
El Salvador	8 Jun	1998	30 Nov	1998
Equatorial Guinea			16 Aug	2000 a
Eritrea			28 Jul	2005 a
Estonia	3 Dec	1998	14 Oct	2002
Eswatini			13 Jan	2006 a
Ethiopia			14 Apr	2005 a

Participant	*Signature*		*Approval(AA), Acceptance(A), Accession(a), Succession(d), Ratification*	
European Union	29 Apr	1998	31 May	2002 AA
Fiji	17 Sep	1998	17 Sep	1998
Finland	29 Apr	1998	31 May	2002
France	29 Apr	1998	31 May	2002 AA
Gabon			12 Dec	2006 a
Gambia			1 Jun	2001 a
Georgia			16 Jun	1999 a
Germany	29 Apr	1998	31 May	2002
Ghana			30 May	2003 a
Greece	29 Apr	1998	31 May	2002
Grenada			6 Aug	2002 a
Guatemala	10 Jul	1998	5 Oct	1999
Guinea			7 Sep	2000 a
Guinea Bissau			18 Nov	2005 a
Guyana			5 Aug	2003 a
Haiti			6 Jul	2005 a
Honduras	25 Feb	1999	19 Jul	2000
Hungary			21 Aug	2002 a
Iceland			23 May	2002 a
India			26 Aug	2002 a
Indonesia	13 Jul	1998	3 Dec	2004
Iran (Islamic Republic of)			22 Aug	2005 a
Iraq			28 Jul	2009 a
Ireland	29 Apr	1998	31 May	2002
Israel	16 Dec	1998	15 Mar	2004
Italy	29 Apr	1998	31 May	2002
Jamaica			28 Jun	1999 a
Japan	28 Apr	1998	4 Jun	2002 A
Jordan			17 Jan	2003 a
Kazakhstan	12 Mar	1999	19 Jun	2009
Kenya			25 Feb	2005 a
Kiribati			7 Sep	2000 a
Kuwait			11 Mar	2005 a

Participant	*Signature*		*Approval(AA), Acceptance(A), Accession(a), Succession(d), Ratification*	
Kyrgyzstan			13 May	2003 a
Lao People's Democratic Republic			6 Feb	2003 a
Latvia	14 Dec	1998	5 Jul	2002
Lebanon			13 Nov	2006 a
Lesotho			6 Sep	2000 a
Liberia			5 Nov	2002 a
Libya			24 Aug	2006 a
Liechtenstein	29 Jun	1998	3 Dec	2004
Lithuania	21 Sep	1998	3 Jan	2003
Luxembourg	29 Apr	1998	31 May	2002
Madagascar			24 Sep	2003 a
Malawi			26 Oct	2001 a
Malaysia	12 Mar	1999	4 Sep	2002
Maldives	16 Mar	1998	30 Dec	1998
Mali	27 Jan	1999	28 Mar	2002
Malta	17 Apr	1998	11 Nov	2001
Marshall Islands	17 Mar	1998	11 Aug	2003
Mauritania			22 Jul	2005 a
Mauritius			9 May	2001 a
Mexico	9 Jun	1998	7 Sep	2000
Micronesia (Federated States of)	17 Mar	1998	21 Jun	1999
Monaco	29 Apr	1998	27 Feb	2006
Mongolia			15 Dec	1999 a
Montenegro			4 Jun	2007 a
Morocco			25 Jan	2002 a
Mozambique			18 Jan	2005 a
Myanmar			13 Aug	2003 a
Namibia			4 Sep	2003 a
Nauru			16 Aug	2001 a
Nepal			16 Sep	2005 a
Netherlands (Kingdom of the)[5]	29 Apr	1998	31 May	2002 A

Participant	***Signature***		***Approval(AA), Acceptance(A), Accession(a), Succession(d), Ratification***	
New Zealand[6]	22 May	1998	19 Dec	2002
Nicaragua	7 Jul	1998	18 Nov	1999
Niger	23 Oct	1998	30 Sep	2004
Nigeria			10 Dec	2004 a
Niue	8 Dec	1998	6 May	1999
North Macedonia			18 Nov	2004 a
Norway	29 Apr	1998	30 May	2002
Oman			19 Jan	2005 a
Pakistan			11 Jan	2005 a
Palau			10 Dec	1999 a
Panama	8 Jun	1998	5 Mar	1999
Papua New Guinea	2 Mar	1999	28 Mar	2002
Paraguay	25 Aug	1998	27 Aug	1999
Peru	Nov	1998	12 Sep	2002
Philippines	15 Apr	1998	20 Nov	2003
Poland	15 Jul	1998	13 Dec	2002
Portugal	29 Apr	1998	31 May	2002 AA
Qatar			11 Jan	2005 a
Republic of Korea	25 Sep	1998	8 Nov	2002
Republic of Moldova			22 Apr	2003 a
Romania	5 Jan	1999	19 Mar	2001
Russian Federation	11 Mar	1999	18 Nov	2004
Rwanda			22 Jul	2004 a
Samoa	16 Mar	1998	27 Nov	2000
San Marino			28 Apr	2010 a
Sao Tome and Principe			25 Apr	2008 a
Saudi Arabia			31 Jan	2005 a
Senegal			20 Jul	2001 a
Serbia			19 Oct	2007 a
Seychelles	20 Mar	1998	22 Jul	2002
Sierra Leone			10 Nov	2006 a
Singapore			12 Apr	2006 a
Slovakia	26 Feb	1999	31 May	2002

Participant	*Signature*		*Approval(AA), Acceptance(A), Accession(a), Succession(d), Ratification*	
Slovenia	21 Oct	1998	2 Aug	2002
Solomon Islands	29 Sep	1998	13 Mar	2003
Somalia			26 Jul	2010 a
South Africa			31 Jul	2002 a
Spain	29 Apr	1998	31 May	2002
Sri Lanka			3 Sep	2002 a
St. Kitts and Nevis			8 Apr	2008 a
St. Lucia	16 Mar	1998	20 Aug	2003
St. Vincent and the Grenadines	19 Mar	1998	31 Dec	2004
Sudan			2 Nov	2004 a
Suriname			25 Sep	2006 a
Sweden	29 Apr	1998	31 May	2002
Switzerland	16 Mar	1998	9 Jul	2003
Syrian Arab Republic			27 Jan	2006 a
Tajikistan			29 Dec	2008 a
Thailand	2 Feb	1999	28 Aug	2002
Timor Leste			14 Oct	2008 a
Togo			2 Jul	2004 a
Tonga			14 Jan	2008 a
Trinidad and Tobago	7 Jan	1999	28 Jan	1999
Tunisia			22 Jan	2003 a
Türkiye			28 May	2009 a
Turkmenistan	28 Sep	1998	11 Jan	1999
Tuvalu Nov		1998	16 Nov	1998
Uganda			25 Mar	2002 a
Ukraine	15 Mar	1999	12 Apr	2004
United Arab Emirates			26 Jan	2005 a
United Kingdom of Great Britain and Northern Ireland,[78]	29 Apr	1998	31 May	2002
United Republic of Tanzania			26 Aug	2002 a

Participant	*Signature*		*Approval(AA), Acceptance(A), Accession(a), Succession(d), Ratification*	
United States of America	12 Nov	1998		
Uruguay	29 Jul	1998	5 Feb	2001
Uzbekistan	Nov	1998	12 Oct	1999
Vanuatu			17 Jul	2001 a
Venezuela (Bolivarian Republic of)			18 Feb	2005 a
Viet Nam	3 Dec	1998	25 Sep	2002
Yemen			15 Sep	2004 a
Zambia	5 Aug	1998	7 Jul	2006
Zimbabwe			30 Jun	2009 a

Declarations and Reservations

(Unless otherwise indicated, the declarations and reservations were made upon ratification, accession, acceptance or approval.)

Australia

"The Government of Australia declares that it is eligible to apply the second sentence of Article 3.7 of the Protocol, using the Revised 1996 IPCC methodologies, as stipulated in Article 5.2 of the Protocol and paragraph 5(b) of the Annex to Decision 13/CMP.1."

Cook Islands

"The Government of the Cook Islands declares its understanding that signature and subsequent ratification of the Kyoto Protocol shall in no way constitute a renunciation of any rights under international law concerning State responsibility for the adverse effects of climate change and that no provision in the Protocol can be interpreted as derogating from principles of general international law.

In this regard, the Government of the Cook Islands further declares that, in light of the best available scientific information and assessment on climate change and its impacts, it considers the emissions reduction obligation in article 3 of the Kyoto Protocol to be inadequate to prevent dangerous anthropogenic interference with the climate system."

European Union

"The European Community and its Member States will fulfil their respective commitments under article 3, paragraph 1, of the Protocol jointly in accordance with the provisions of article 4."

Declaration by the European Community made in accordance with article 24(3) of the Kyoto Protocol.

"The following States are at present members of the European Community: the Kingdom of Belgium, the Kingdom of Denmark, the Federal Republic of Germany, the Hellenic Republic, the Kingdom of Spain, the French Republic, Ireland, the Italian Republic, the Grand Duchy of Luxembourg, the Kingdom of the Netherlands, the Republic of Austria, the Portuguese Republic, the Republic of Finland, the Kingdom of Sweden, the United Kingdom of Great Britain and Northern Ireland.

The European Community declares that, in accordance with the Treaty establishing the European Community, and in particular article 175(1) thereof, it is competent to enter into international agreements, and to implement the obligations resulting therefrom, which contribute to the pursuit of the following objectives:

- preserving, protecting and improving the quality of the environment;
- protecting human health;
- prudent and rational utilisation of natural resources;
- promoting measures at international level to deal with regional or world wide environmental problems.

The European Community declares that its quantified emission reduction commitment under the Protocol will be fulfilled through action by the Community and its Member States within the respective competence of each and that it has already adopted legal instruments, binding on its Member States, covering matters governed by the Protocol.

The European Community will on a regular basis provide information on relevant Community legal instruments within the framework of the supplementary information incorporated in its national communication submitted under art 12 of the Convention for the purpose of demonstrating compliance with its commitments under the Protocol in accordance with article 7(2) thereof and the guidelines thereunder."

Ireland

"The European Community and the Member States, including Ireland, will fulfil their respective commitments under article 3, paragraph 1, of the Protocol in accordance with the provisions of article 4."

Kiribati

"The Government of the Republic of Kiribati declares its understanding that accession to the Kyoto Protocol shall in no way constitute a renunciation of any rights under international law concerning State responsibility for the adverse effects of the climate change and that no provision in the Protocol can be interpreted as derogating from principles of general international law."

Nauru

"...The Government of the Republic of Nauru declares its understanding that the ratification of the Kyoto Protocol shall in no way constitute a renunciation of any rights under international law concerning State responsibility for the adverse effects of climate change; ...

... The Government of the Republic of Nauru further declares that, in the light of the best available scientific information and assessment of climate change and impacts, it considers the emissions of reduction obligations in Article 3 of the Kyoto Protocol to be inadequate to prevent the dangerous anthropogenic interference with the climate system;

... [The Government of the Republic of Nauru declares] that no provisions in the Protocol can be interpreted as derogating from the principles of general international law[.]"

Niue

"The Government of Niue declares its understanding that ratification of the Kyoto Protocol shall in no way constitute a renunciation of any rights under international law concerning state responsibility for the adverse effects of climate change and that no provisions in the Protocol can be interpreted as derogating from the principles of general international law.

In this regard, the Government of Niue further declares that, in light of the best available scientific information and assessment of climate change and impacts, it considers the emissions reduction obligations in article 3 of the Kyoto Protocol to be inadequate to prevent dangerous anthropogenic interference with the climate system."

Russian Federation

The Russian Federation proceeds from the assumption that the commitments of the Russian Federation under the Protocol will have serious consequences for its social and economic development. Therefore, the decision on ratification was taken following a thorough analysis of all factors, inter alia, the importance of the Protocol for the promotion of international cooperation, and taking into account that the Protocol can enter into force only if the Russian Federation ratifies it.

The Protocol establishes for each of the Parties that have signed it quantified reductions of greenhouse gas emissions to atmosphere for the first commitment period from 2008 to 2012.

The commitments of the Parties to the Protocol on quantified reductions of greenhouse gas emissions to atmosphere for the second and subsequent commitment periods of the Protocol, that is after 2012, will be established through negotiations of the Parties to the Protocol scheduled to start in 2005. On the outcome of these negotiations the Russian Federation will take a decision on its participation in the Protocol in the second and subsequent commitment periods.

Syrian Arab Republic

The accession of the Syrian Arab Republic to this Protocol shall in no way imply its recognition of Israel or entail its entry into any dealings with Israel in the matters governed by the provisions thereof.

Notes

1. For the purpose of entry into force of the [Convention/Protocol] any instrument of ratification, acceptance, approval or accession deposited by a regional economic integration organization shall not be counted as additional to those deposited by member States of that Organization.
2. In accordance with article 27(2) of the Kyoto Protocol to the United Nations Framework Convention on Climate Change, the Government of Canada notified the Secretary-General that it had decided to withdraw from the Kyoto Protocol as from the date indicated hereinafter:

Participant:	*Date of notification*:	*Date of effect*:
Canada	15 Dec 2011	15 Dec 2012

3. In a communication received on 30 August 2002, the Government of the People's Republic of China informed the Secretary-General of the following:

   In accordance with article 153 of the Basic Law of the Hong Kong Special Administrative Region of the People's Republic of China of 1990 and article 138 of the Basic Law of the Macao Special Administrative Region of the People's Republic of China of 1993, the Government of the People's Republic of China decides that the Kyoto Protocol to the United Nations Framework Convention on Climate Change shall provisionally not apply to the Hong Kong Special Administrative Region and the Macao Special Administrative Region of the People's Republic of China.

   Further, in a communication received on 8 April 2003, the Government of the Government of the People's Republic of China notified the Secretary-General of the following:

   "In accordance with the provisions of Article 153 of the Basic Law of the Hong Kong Special Administrative Region of the People's Republic of China of 1990, the Government of the People's Republic of China decides that the United Nations Framework Convention on Climate Change and the Kyoto Protocol to the United Nations Framework Convention on Climate Change shall apply to the Hong Kong Special Administrative Region of the People's Republic of China.

   The United Nations Framework Convention on Climate Change continues to be implemented in the Macao Special Administrative Region of the People's Republic of China. The Kyoto Protocol to the United Nations Framework Convention on Climate Change shall not apply to the Macao Special Administrative Region of the People's Republic of China until the Government of China notifies otherwise."

   In a communication received on 14 January 2008, the Government of the Government of the People's Republic of China notified the Secretary-General of the following:

   In accordance with Article 138 of the Basic Law of the Macao Special Administrative Region of the People's Republic of China, the Government of the People's Republic of China decides that the Kyoto Protocol to the United Nations Framework Convention on Climate Change shall apply to the Macao Special Administrative Region of the People's Republic of China.
4. With a territorial exclusion to the Faroe Islands.
5. For the Kingdom in Europe.
6. With the following declaration:

   "…..consistent with the constitutional status of Tokelau and taking into account the commitment of the Government of New Zealand to the development of self-government for Tokelau through an act of self-determination under the Charter of the United Nations, this ratification shall not extend to Tokelau unless and until a Declaration to this effect is lodged by the Government of New

Zealand with the Depositary on the basis of appropriate consultation with that territory."

7. By a communication received on 27 March 2007, the Government of Argentina notified the Secretary-General of the following:

   The Argentine Republic objects to the extension of the territorial application to the Kyoto Protocol to the United Nations Framework Convention on Climate Change of 11 December 1997 with respect to the Malvinas Islands, which was notified by the United Kingdom of Great Britain and Northern Ireland to the Depositary of the Convention on 7 March 2007.

   The Argentine Republic reaffirms its sovereignty over the Malvinas Islands, the South Georgia and South Sandwich Islands and the surrounding maritime spaces, which are an integral part of its national territory, and recalls that the General Assembly of the United Nations adopted resolutions 2065 (XX), 3160 (XXVIII), 31/49, 37/9, 38/12, 39/6, 40/21, 41/40, 42/19 and 43/25, which recognize the existence of a dispute over sovereignty and request the Governments of the Argentine Republic and the United Kingdom of Great Britain and Northern Ireland to initiate negotiations with a view to finding the means to resolve peacefully and definitively the pending problems between both countries, including all aspects on the future of the Malvinas Islands, in accordance with the Charter of the United Nations.

8. On 4 April 2006, the Government of the United Kingdom informed the Secretary-General that the Protocol shall apply to the Bailiwick of Guernsey and the Isle of Man. On 2 January 2007: in respect of Gibraltar. On 7 March 2007: in respect of Bermuda, Cayman Islands, Falkland Islands (Malvinas) and the Bailiwick of Jersey.

Appendix 3

Paris Agreement

Paris, 12 December 2015

ENTRY INTO FORCE:	4 November 2016, in accordance with article 21(1). The Agreement enters into force on the thirtieth day after the date on which at least 55 Parties to the Convention accounting in total for at least an estimated 55 per cent of the total global greenhouse gas emissions have deposited their instruments of ratification, acceptance, approval or accession
REGISTRATION:	4 November 2016, No. 54113
STATUS:	Signatories: 195. Parties: 195
TEXT:	United Nations, *Treaty Series*, vol. 3156, p. 79 C.N.63.2016.TREATIES-XXVII.7.d of 16 February 2016 (Opening for signature) and C.N.92.2016.TREATIES-XXVII.7.d of 17 March 2016 (Issuance of Certified True Copies)

Note: The Paris Agreement was adopted on 12 December 2015 at the twenty-first session of the Conference of the Parties to the United Nations Framework Convention on Climate Change held in Paris from 30 November to 13 December 2015. In accordance with its article 20, the Agreement shall be open for signature at the United Nations Headquarters in New York from 22 April 2016 until 21 April 2017 by States and regional economic integration organizations that are Parties to the United Nations Framework Convention on Climate Change.

Participant	*Signature*		*Approval(AA), Acceptance(A), Accession(a), Succession(d), Ratification*	
Afghanistan	22 Apr	2016	15 Feb	2017

X. Zou, *Thirty Years of Global Climate Governance*, Energy, Climate Change & Sustainability in the Global South, https://doi.org/10.1007/978-981-95-8457-4

Participant	*Signature*		*Approval(AA), Acceptance(A), Accession(a), Succession(d), Ratification*	
Albania	22 Apr	2016	21 Sep	2016
Algeria	22 Apr	2016	20 Oct	2016
Andorra	22 Apr	2016	24 Mar	2017
Angola	22 Apr	2016	16 Nov	2020
Antigua and Barbuda	22 Apr	2016	21 Sep	2016
Argentina	22 Apr	2016	21 Sep	2016
Armenia	20 Sep	2016	23 Mar	2017
Australia	22 Apr	2016	9 Nov	2016
Austria	22 Apr	2016	5 Oct	2016
Azerbaijan	22 Apr	2016	9 Jan	2017
Bahamas (The)	22 Apr	2016	22 Aug	2016
Bahrain	22 Apr	2016	23 Dec	2016
Bangladesh	22 Apr	2016	21 Sep	2016
Barbados	22 Apr	2016	22 Apr	2016
Belarus	22 Apr	2016	21 Sep	2016 A
Belgium	22 Apr	2016	6 Apr	2017
Belize	22 Apr	2016	22 Apr	2016
Benin	22 Apr	2016	31 Oct	2016
Bhutan	22 Apr	2016	19 Sep	2017
Bolivia (Plurinational State of)	22 Apr	2016	5 Oct	2016
Bosnia and Herzegovina	22 Apr	2016	16 Mar	2017
Botswana	22 Apr	2016	11 Nov	2016
Brazil	22 Apr	2016	21 Sep	2016
Brunei Darussalam	22 Apr	2016	21 Sep	2016
Bulgaria	22 Apr	2016	29 Nov	2016
Burkina Faso	22 Apr	2016	11 Nov	2016
Burundi	22 Apr	2016	17 Jan	2018
Cabo Verde	22 Apr	2016	21 Sep	2017
Cambodia	22 Apr	2016	6 Feb	2017
Cameroon	22 Apr	2016	29 Jul	2016
Canada	22 Apr	2016	5 Oct	2016

Participant	*Signature*		*Approval(AA), Acceptance(A), Accession(a), Succession(d), Ratification*	
Central African Republic	22 Apr	2016	11 Oct	2016
Chad	22 Apr	2016	12 Jan	2017
Chile	20 Sep	2016	10 Feb	2017
China	22 Apr	2016	3 Sep	2016
Colombia	22 Apr	2016	12 Jul	2018
Comoros	22 Apr	2016	23 Nov	2016
Congo	22 Apr	2016	21 Apr	2017
Cook Islands	24 Jun	2016	1 Sep	2016
Costa Rica	22 Apr	2016	13 Oct	2016
Côte d'Ivoire	22 Apr	2016	25 Oct	2016
Croatia	22 Apr	2016	24 May	2017
Cuba	22 Apr	2016	28 Dec	2016
Cyprus	22 Apr	2016	4 Jan	2017
Czech Republic	22 Apr	2016	5 Oct	2017
Democratic People's Republic of Korea	22 Apr	2016	1 Aug	2016
Democratic Republic of the Congo	22 Apr	2016	13 Dec	2017
Denmark[1]	22 Apr	2016	1 Nov	2016 AA
Djibouti	22 Apr	2016	11 Nov	2016
Dominica	22 Apr	2016	21 Sep	2016
Dominican Republic	22 Apr	2016	21 Sep	2017
Ecuador	26 Jul	2016	20 Sep	2017
Egypt	22 Apr	2016	29 Jun	2017
El Salvador	22 Apr	2016	27 Mar	2017
Equatorial Guinea	22 Apr	2016	30 Oct	2018
Eritrea	22 Apr	2016	7 Feb	2023
Estonia	22 Apr	2016	4 Nov	2016
Eswatini	22 Apr	2016	21 Sep	2016
Ethiopia	22 Apr	2016	9 Mar	2017
European Union	22 Apr	2016	5 Oct	2016
Fiji	22 Apr	2016	22 Apr	2016
Finland	22 Apr	2016	14 Nov	2016
France	22 Apr	2016	5 Oct	2016

Participant	*Signature*		*Approval(AA), Acceptance(A), Accession(a), Succession(d), Ratification*	
Gabon	22 Apr	2016	2 Nov	2016
Gambia	26 Apr	2016	7 Nov	2016
Georgia	22 Apr	2016	8 May	2017 AA
Germany	22 Apr	2016	5 Oct	2016
Ghana	22 Apr	2016	21 Sep	2016
Greece	22 Apr	2016	14 Oct	2016
Grenada	22 Apr	2016	22 Apr	2016
Guatemala	22 Apr	2016	25 Jan	2017
Guinea	22 Apr	2016	21 Sep	2016
Guinea Bissau	22 Apr	2016	22 Oct	2018
Guyana	22 Apr	2016	20 May	2016
Haiti	22 Apr	2016	31 Jul	2017
Holy See			4 Sep	2022 a
Honduras	22 Apr	2016	21 Sep	2016
Hungary	22 Apr	2016	5 Oct	2016
Iceland	22 Apr	2016	21 Sep	2016 A
India	22 Apr	2016	2 Oct	2016
Indonesia	22 Apr	2016	31 Oct	2016
Iran (Islamic Republic of)	22 Apr	2016		
Iraq	8 Dec	2016	1 Nov	2021
Ireland	22 Apr	2016	4 Nov	2016
Israel	22 Apr	2016	22 Nov	2016
Italy	22 Apr	2016	11 Nov	2016
Jamaica	22 Apr	2016	10 Apr	2017
Japan	22 Apr	2016	8 Nov	2016 A
Jordan	22 Apr	2016	4 Nov	2016
Kazakhstan	2 Aug	2016	6 Dec	2016
Kenya	22 Apr	2016	28 Dec	2016
Kiribati	22 Apr	2016	21 Sep	2016
Kuwait	22 Apr	2016	23 Apr	2018
Kyrgyzstan	21 Sep	2016	18 Feb	2020
Lao People's Democratic Republic	22 Apr	2016	7 Sep	2016
Latvia	22 Apr	2016	16 Mar	2017

Participant	*Signature*		*Approval(AA), Acceptance(A), Accession(a), Succession(d), Ratification*	
Lebanon	22 Apr	2016	5 Feb	2020
Lesotho	22 Apr	2016	20 Jan	2017
Liberia	22 Apr	2016	27 Aug	2018
Libya	22 Apr	2016		
Liechtenstein	22 Apr	2016	20 Sep	2017
Lithuania	22 Apr	2016	2 Feb	2017
Luxembourg	22 Apr	2016	4 Nov	2016
Madagascar	22 Apr	2016	21 Sep	2016
Malawi	20 Sep	2016	29 Jun	2017
Malaysia	22 Apr	2016	16 Nov	2016
Maldives	22 Apr	2016	22 Apr	2016
Mali	22 Apr	2016	23 Sep	2016
Malta	22 Apr	2016	5 Oct	2016
Marshall Islands	22 Apr	2016	22 Apr	2016
Mauritania	22 Apr	2016	27 Feb	2017
Mauritius	22 Apr	2016	22 Apr	2016
Mexico	22 Apr	2016	21 Sep	2016
Micronesia (Federated States of)	22 Apr	2016	15 Sep	2016
Monaco	22 Apr	2016	24 Oct	2016
Mongolia	22 Apr	2016	21 Sep	2016
Montenegro	22 Apr	2016	20 Dec	2017
Morocco	22 Apr	2016	21 Sep	2016
Mozambique	22 Apr	2016	4 Jun	2018
Myanmar	22 Apr	2016	19 Sep	2017
Namibia	22 Apr	2016	21 Sep	2016
Nauru	22 Apr	2016	22 Apr	2016
Nepal	22 Apr	2016	5 Oct	2016
Netherlands (Kingdom of the)[2]	22 Apr	2016	28 Jul	2017 A
New Zealand[3]	22 Apr	2016	4 Oct	2016
Nicaragua			23 Oct	2017 a
Niger	22 Apr	2016	21 Sep	2016
Nigeria	22 Apr	2016	16 May	2017
Niue	28 Oct	2016	28 Oct	2016

Participant	*Signature*		*Approval(AA), Acceptance(A), Accession(a), Succession(d), Ratification*	
North Macedonia	22 Apr	2016	9 Jan	2018
Norway	22 Apr	2016	20 Jun	2016
Oman	22 Apr	2016	22 May	2019
Pakistan	22 Apr	2016	10 Nov	2016
Palau	22 Apr	2016	22 Apr	2016
Panama	22 Apr	2016	21 Sep	2016
Papua New Guinea	22 Apr	2016	21 Sep	2016
Paraguay	22 Apr	2016	14 Oct	2016
Peru	22 Apr	2016	25 Jul	2016
Philippines	22 Apr	2016	23 Mar	2017
Poland	22 Apr	2016	7 Oct	2016
Portugal	22 Apr	2016	5 Oct	2016
Qatar	22 Apr	2016	23 Jun	2017
Republic of Korea	22 Apr	2016	3 Jun	2016
Republic of Moldova	22 Apr	2016	20 Jun	2017
Romania	22 Apr	2016	1 Jun	2017
Russian Federation	22 Apr	2016	7 Oct	2019 A
Rwanda	22 Apr	2016	6 Oct	2016
Samoa	22 Apr	2016	22 Apr	2016
San Marino	22 Apr	2016	26 Sep	2018
Sao Tome and Principe	22 Apr	2016	2 Nov	2016
Saudi Arabia	Nov	2016	3 Nov	2016
Senegal	22 Apr	2016	21 Sep	2016
Serbia	22 Apr	2016	25 Jul	2017
Seychelles	25 Apr	2016	29 Apr	2016
Sierra Leone	22 Sep	2016	1 Nov	2016
Singapore	22 Apr	2016	21 Sep	2016
Slovakia	22 Apr	2016	5 Oct	2016
Slovenia	22 Apr	2016	16 Dec	2016
Solomon Islands	22 Apr	2016	21 Sep	2016
Somalia	22 Apr	2016	22 Apr	2016
South Africa	22 Apr	2016	1 Nov	2016
South Sudan	22 Apr	2016	23 Feb	2021

Participant	*Signature*		*Approval(AA), Acceptance(A), Accession(a), Succession(d), Ratification*	
Spain	22 Apr	2016	12 Jan	2017
Sri Lanka	22 Apr	2016	21 Sep	2016
St. Kitts and Nevis	22 Apr	2016	22 Apr	2016
St. Lucia	22 Apr	2016	22 Apr	2016
St. Vincent and the Grenadines	22 Apr	2016	29 Jun	2016
State of Palestine	22 Apr	2016	22 Apr	2016
Sudan	22 Apr	2016	2 Aug	2017
Suriname	22 Apr	2016	13 Feb	2019
Sweden	22 Apr	2016	13 Oct	2016
Switzerland	22 Apr	2016	6 Oct	2017
Syrian Arab Republic			13 Nov	2017 a
Tajikistan	22 Apr	2016	22 Mar	2017
Thailand	22 Apr	2016	21 Sep	2016
Timor Leste	22 Apr	2016	16 Aug	2017
Togo	19 Sep	2016	28 Jun	2017
Tonga	22 Apr	2016	21 Sep	2016
Trinidad and Tobago	22 Apr	2016	22 Feb	2018
Tunisia	22 Apr	2016	10 Feb	2017
Türkiye	22 Apr	2016	11 Oct	2021
Turkmenistan	23 Sep	2016	20 Oct	2016
Tuvalu	22 Apr	2016	22 Apr	2016
Uganda	22 Apr	2016	21 Sep	2016
Ukraine	22 Apr	2016	19 Sep	2016
United Arab Emirates	22 Apr	2016	21 Sep	2016 A
United Kingdom of Great Britain and Northern Ireland,[4],[5][6]	22 Apr	2016	18 Nov	2016
United Republic of Tanzania	22 Apr	2016	18 May	2018
United States of America[7]			20 Jan	2021 A
Uruguay	22 Apr	2016	19 Oct	2016
Uzbekistan	19 Apr	2017	9 Nov	2018

Participant	*Signature*		*Approval(AA), Acceptance(A), Accession(a), Succession(d), Ratification*	
Vanuatu	22 Apr	2016	21 Sep	2016
Venezuela (Bolivarian Republic of)		2016	21 Jul	2017
Viet Nam	22 Apr	2016	3 Nov	2016 AA
Yemen	23 Sep	2016		
Zambia	20 Sep	2016	9 Dec	2016
Zimbabwe	22 Apr	2016	7 Aug	2017

Declarations

(Unless otherwise indicated, the declarations were made upon ratification, acceptance, approval or accession.)

Belgium

"This signature engages also the Walloon Region, the Flemish Region and the Brussels-Capital Region."

Bulgaria

"The Republic of Bulgaria recognizes that in accordance with Article 9, paragraph 1, of the Paris Agreement developed country Parties shall provide financial resources to assist developing country Parties with respect to both mitigation and adaptation in continuation of their existing obligations under the Convention. In this context the Republic of Bulgaria notes that as a Party to the United Nations Framework Convention on Climate Change Bulgaria is not included in Annex II."

China

In accordance with the Basic Law of the Hong Kong Special Administrative Region of the People's Republic of China and the Basic Law of the Macao Special Administrative Region of the People's Republic of China, the Government of the People's Republic of China decides that the Agreement applies to the Hong Kong Special Administrative Region and the Macao Special Administrative Region of the People's Republic of China.

Cook Islands

The Government of the Cook Islands declares its understanding that acceptance of the Paris Agreement and its application shall in no way constitute a renunciation of any rights under international law concerning State responsibility for the adverse effects of climate change and that no provision in the Paris Agreement can be interpreted as derogating from principles of general international law or any claims or rights concerning compensation due to the impacts of climate change.

The Government of the Cook Islands further declares that, in light of the best available scientific information and assessment on climate change and its impacts, it considers the emissions reduction obligations in the aforesaid Paris Agreement to be inadequate to prevent a global temperature stabilisation level at or above 1.5 degrees Celsius relative to pre-industrial levels and as a consequence, such emissions will have severe implications for our national interests.

European Union

"Declaration by the Union made in accordance with Article 20(3) of the Paris Agreement.

The following States are at present Members of the European Union: the Kingdom of Belgium, the Republic of Bulgaria, the Czech Republic, the Kingdom of Denmark, the Federal Republic of Germany, the Republic of Estonia, Ireland, the Hellenic Republic, the Kingdom of Spain, the French Republic, the Republic of Croatia, the Italian Republic, the Republic of Cyprus, the Republic of Latvia, the Republic of Lithuania, the Grand Duchy of Luxembourg, Hungary, the Republic of Malta, the Kingdom of the Netherlands, the Republic of Austria, the Republic of Poland, the Portuguese Republic, Romania, the Republic of Slovenia, the Slovak Republic, the Republic of Finland, the Kingdom of Sweden, the United Kingdom of Great Britain and Northern Ireland.

The European Union declares that, in accordance with the Treaty on the Functioning of the European Union, and in particular Article 191 and Article 192(1) thereof, it is competent to enter into international agreements, and to implement the obligations resulting therefrom, which contribute to the pursuit of the following objectives:

- preserving, protecting and improving the quality of the environment;
- protecting human health;
- prudent and rational utilisation of natural resources;
- promoting measures at international level to deal with regional or worldwide environmental problems, and in particular combating climate change.

…

The European Union will continue to provide information, on a regular basis on any substantial modifications in the extent of its competence, in accordance with Article 20(3) of the Agreement."

Holy See

"By acceding to the Paris Agreement under the United Nations Framework Convention on Climate Change in the name and on behalf of Vatican City State, the Holy See intends to contribute to the efforts of all States to work together in solidarity, in accordance with their common but differentiated responsibilities and respective capabilities, in an effective response to the challenges posed by climate change to humankind and to our common home.

In light of the territorial nature of the obligations set forth in the Paris Agreement, the Holy See declares, for the avoidance of doubt, that in acceding to the Agreement only in the name and on behalf of Vatican City State it commits itself to apply its provisions exclusively within the Territory of the Vatican City State, as circumscribed by the Leonine Walls.

The Holy See, in conformity with its particular mission, reiterates, on behalf of Vatican City State, its position regarding the term 'gender' used in the Preamble (PP11) and articles 7.5 and 11.2 of the Paris Agreement. The Holy See underlines that any reference to 'gender' and related terms in any document that has been or that will be adopted by the Conference of State Parties or by its subsidiary bodies is to be understood as grounded on the biological sexual identity that is male and female.

The Holy See upholds and promotes a holistic and integrated approach that is firmly centered on the human dignity and integral development of every person."

India

"The Government of India declares its understanding that, as per its national laws; keeping in view its development agenda, particularly the eradication of poverty and provision of basic needs for all its citizens, coupled with its commitment to following the low carbon path to progress, and on the assumption of unencumbered availability of cleaner sources of energy and technologies and financial resources from around the world; and based on a fair and ambitious assessment of global commitment to combating climate change, it is ratifying the Paris Agreement."

Marshall Islands

"...the Government of the Republic of the Marshall Islands declares its understanding that ratification of the Paris Agreement shall in no way constitute a renunciation of any rights under any other laws, including international law, and the communication depositing the Republic's instrument of ratification shall include a declaration to this effect for international record;

FURTHERMORE, the Government of the Republic of the Marshall Islands declares that, in light of best scientific information and assessment on climate change and its impacts, it considers the emission reduction obligations in Article 3 of the Kyoto Protocol, the Doha Amendment and the aforesaid Paris Agreement to be inadequate to prevent global temperature increase of 1.5 degrees Celsius above pre-Industrial levels and as a consequence, will have severe implications for our national interests..."

Mexico

... in accordance with their national legal framework, and in consideration of the best and most up-to-date scientific information available and incorporated by the Intergovernmental Panel on Climate Change, the United Mexican States understands greenhouse gas emissions to mean the release into the atmosphere of greenhouse gases and/or their precursors and aerosols into the atmosphere, including, where applicable, greenhouse compounds, within a specific area and during a specific period of time.

Micronesia (Federated States of)

"The Government of the Federated States of Micronesia declares its understanding that its ratification of the Paris Agreement does not constitute a renunciation of any rights of the Government of the Federated States of Micronesia under international law concerning State responsibility for the adverse effects of climate change, and that no provision in the Paris Agreement can be interpreted as derogating from principles of general international law or any claims or rights concerning compensation and liability due to the adverse effects of climate change; and

The Government of the Federated States of Micronesia further declares that, in light of the best available scientific information and assessments on climate change and its impacts, it considers the emission reduction obligations in the Paris Agreement to be inadequate to prevent a global temperature increase above 1.5 degrees Celsius relative to pre-industrial levels, and as a consequence, such emissions will have severe implications for the national interests of the Government of the Federated States of Micronesia."

Nauru

"... the Government, of Nauru declares its understanding that the ratification of the Agreement shall in no way constitute a renunciation of any rights under international law concerning State responsibility [for] the adverse effects of climate change.

FURTHER, the Government of Nauru declares that no provisions in the Agreement can be interpreted as derogating from the principles of general international law.

AND FURTHER, the Government of Nauru declares its understanding that Article 8 and decision 1/CP.21, paragraph 51 in no way limits the ability of Parties to UNFCCC or the Agreement to raise, discuss, or address any present or future concerns regarding the issues of liability and compensation.

The Republic of Nauru put forth its concern intended to recognize and acknowledge its national interest..."

Netherlands (Kingdom of The)

"The Kingdom of the Netherlands, for the European part of the Netherlands, declares in accordance with Article 14, paragraph 2, of the United Nations Framework Convention on Climate Change in conjunction with Article 24 of the Paris Agreement, that it accepts both means of dispute settlement referred to in that paragraph as compulsory in relation to any Party accepting one or both means of dispute settlement."

Niue

"The Government of Niue declares its understanding that acceptance of the Paris Agreement and its application shall in no way constitute a renunciation of any rights under international law concerning State responsibility for the adverse effects of climate change and that no provision in the Paris Agreement can be interpreted as derogating from principles of general international law or any claims or rights concerning compensation due to the impacts of climate change.

The Government of Niue further declares that, in light of the best available scientific information and assessment on climate change and its impacts, it considers the emissions reduction obligations in the aforesaid Paris Agreement to be inadequate to prevent a global temperature stabilisation level at or above 1.5 degrees Celsius relative to pre-industrial levels and as a consequence, such emissions will have severe implications for our national interests."

Philippines

"THAT it is the understanding of the Government of the Republic of the Philippines that its accession to and the implementation of the Paris Agreement shall in no way constitute a renunciation of rights under any local and international laws or treaties, including those concerning State responsibility for loss and damage associated with the adverse effects of climate change;

THAT, the accession to and implementation of the Paris Agreement by the Republic of the Philippines is for the purpose of supporting the country's national development objectives and priorities such as sustainable industrial development, the eradication of poverty and provision of basic needs, and securing social and climate justice and energy security for all its citizens."

Poland

"The Government of the Republic of Poland recognizes that under Article 9 paragraph 1 of the Paris Agreement developed country Parties shall provide financial resources to assist developing country Parties with respect to both mitigation and adaptation in continuation of their existing obligations under the Convention. In this context the Government of the Republic of Poland notes that Poland is a Party to the United Nations Framework Convention on Climate Change not included in Annex II."

Russian Federation

1. The Russian Federation recognizes that, in accordance with paragraph 1 of Article 9 of the Agreement, developed country Parties shall provide financial resources to assist developing country Parties with respect to both mitigation of climate change and adaptation to it in continuation of their existing obligations under the United Nations Framework Convention on Climate Change of 9 May 1992 (hereinafter referred to as "the Convention"). In this context, the Russian Federation notes

that as a Party to the Convention the Russian Federation is not included in Annex II to the Convention.

2. The Russian Federation proceeds from the importance of conservation and enhancement of absorbing capacity of forests and other ecosystems, as well as from the necessity of the maximum possible account of this capacity including in the implementation of the Agreement's mechanisms.
3. The Russian Federation considers unacceptable the use of the Agreement and its mechanisms as tools to create barriers to sustainable social and economic development of the Parties to the Convention.

Solomon Islands

"… the Government of Solomon Islands declares its understanding that acceptance of the aforesaid Paris Agreement shall in no way constitute a renunciation of any rights under international law concerning State responsibility for the adverse effects of climate change;

FURTHER, that the Government of Solomon Islands declares that no provision in this Paris Agreement can be interpreted as derogating from principles of general international law or any claims or rights concerning compensation due to impacts of climate change;

AND that the Government of Solomon Islands declares that the low ambition of the Paris Agreement and its adequacy to stabilize global temperature to safe level of below 1.5 degree Celsius, such emissions will have severe impacts and undermining our sustainable development efforts…"

Spain

In the case where this Agreement is ratified by the United Kingdom and its application extended to the territory of Gibraltar, Spain wishes to make the following declaration:

1. Gibraltar is a non-autonomous territory whose international relations come under the responsibility of the United Kingdom and which is subject to a decolonisation process in accordance with the relevant decisions and resolutions of the General Assembly of the United Nations.
2. The authorities of Gibraltar have a local character and exercise exclusively internal competences which have their origin and their foundation in the distribution and attribution of competences performed by the United Kingdom in compliance with its internal legislation, in its capacity as sovereign State on which the mentioned non-autonomous territory depends.

3. As a result, the eventual participation of authorities of Gibraltar in the application of this Agreement will be understood as carried out exclusively as part of the internal competences of Gibraltar and cannot be considered to modify in any way what was established in the two previous paragraphs.
4. The application of this Agreement to Gibraltar cannot be interpreted as an recognition of any rights or situations regarding areas not covered by article 10 of the Treaty of Utrecht of 13 July 1713, concluded between the Crowns of Spain and of the United Kingdom.

Türkiye

The Republic of Turkey, on the basis of "equity, common but differentiated responsibilities and respective capabilities" as clearly and accurately recognized under the United Nations Framework Convention on Climate Change of 9 May 1992 and the Paris Agreement, and by recalling decisions 26/CP.7, 1/CP.16, 2/CP.17, 1/CP.18 and 21/CP.20 adopted by the Conference of the Parties to the Convention, declares that Turkey will implement the Paris Agreement as a developing country and in the scope of her nationally determined contribution statements, provided that the Agreement and its mechanisms do not prejudice her right to economic and social development.

Tuvalu

"The Government of Tuvalu hereby notifies that it will apply the Paris Agreement provisionally as provided for in paragraph 4 of Decision 1/CP.21.

[…]

The Government of Tuvalu further declares its understanding that acceptance of the aforesaid Paris Agreement and its provisional application shall in no way constitute a renunciation of any rights under international law concerning State responsibility for the adverse effects of climate change and that no provision in the Paris Agreement can be interpreted as derogating from principles of general international law or any claims or rights concerning compensation due to the impacts of climate change.

The Government of Tuvalu further declares that, in light of the best available scientific information and assessment on climate change and its impacts, it considers the emissions reduction obligations in the aforesaid Paris Agreement to be inadequate to prevent a global temperature stabilisation level at or above 1.5 degrees Celsius relative to pre-industrial levels and as a consequence, such emissions will have severe implications for our national interests."

Vanuatu

"WHEREAS the Government of the Republic of Vanuatu declares its understanding that ratification of the Paris Agreement shall in no way constitute a renunciation of any rights under any other laws, including international law, and the communication depositing the Republic's instrument of ratification shall include a declaration to this effect for international record;

FURTHERMORE, that the Government of the Republic of Vanuatu declares that, in light of best scientific information and assessment on climate change and its impacts, it considers the emission reduction obligations in Article 3 of the Kyoto Protocol, the Doha Amendment and the aforesaid Paris Agreement to be inadequate to prevent global temperature increase of 1.5 degrees Celsius above pre-Industrial levels and as a consequence, will have severe implications for our national interests..."

Notes

1. By a communication received on 2 July 2024, the Kingdom of Denmark notified the Secretary General that it had decided to withdraw the declaration, made upon approval, regarding the territorial exclusion in respect of Greenland.

   With territorial exclusion in respect of Greenland. See C.N.819.2016.TREATIES-XXVII.7.d of 1 November 2016.
2. For the European Part of the Netherlands.
3. On 13 November 2017, New Zealand notified the Secretary General of the extention of the application of the Agreement to Tokelau (See CN.705.2017.TREATIES-XXVII.7.d of 13 November 2017). Upon ratification on 4 October 2017, New Zealand notified the Secretary General of a territorial exclusion in respect of Tokelau (See C.N.723.2016.TREATIES-XXVII.7.d of 4 October 2016).
4. On 23 Septembre 2022, the Government of the United Kingdom of Great Britain and Northern Ireland notified the Secretary-General that its ratification of the Agreement would extend to the territory of Gilbratar as follows:

   "... the Government of the United Kingdom of Great Britain and Northern Ireland hereby extends the application of the United Kingdom's ratification of the Agreement to the territory of Gibraltar, for the international relations of which the United Kingdom is responsible.

   The Government of the United Kingdom of Great Britain and Northern Ireland considers the extension of the Agreement to the territory of Gilbratar to take effect on the date of deposit of this notification".

   On 29 April 2022, the Government of the United Kingdom of Great Britain and Northern Ireland notified the Secretary-General that its ratification of the Agreement would extend to the territory of the Bailiwick of Jersey as follows:

   "... the Government of the United Kingdom of Great Britain and Northern Ireland hereby extends the application of the United Kingdom's ratification of the Agreement to the territory of the Bailiwick of Jersey, for the international relations of which the United Kingdom is responsible.

The Government of the United Kingdom of Great Britain and Northern Ireland considers the extension of the Agreement to the territory of the Bailiwick of Jersey to take effect on the date of deposit of this notification".

5. On 22 March 2023, the Government of the United Kingdom of Great Britain and Northern Ireland notified the Secretary-General that its ratification of the Agreement would extend to the Crown Dependency of the Isle of Man as follows:

   "... the Government of the United Kingdom of Great Britain and Northern Ireland hereby extends the application of the United Kingdom's ratification of the [Paris Agreement] to the Crown Dependency of the Isle of Man, a territory for the international relations of which the United Kingdom of Great Britain and Northern Ireland is responsible.

   The Government of the United Kingdom of Great Britain and Northern Ireland considers the extension of the [Paris Agreement] to the Crown Dependency of the Isle of Man to take effect on the date of deposit of this notification."

6. On 6 December 2023, the Government of the United Kingdom of Great Britain and Northern Island notified the Secretary-General of the extension of the application of the Agreement to the territory of Guernsey.

   (See depositary notification C.N.512.2023.TREATIES-XXVII.7.d of 18 December 2023).

7. On 3 September 2016, the Government of the United States of America deposited its instrument of acceptance of the Agreement. See C.N.612.2016.TREATIES-XXVII.7.d of 3 September 2016. On 4 November 2019, the Government of the United States of America notified the Secretary-General of its decision to withdraw from the Agreement which took effect on 4 November 2020 in accordance with article 28(1) and (2) of the Agreement. See C.N.575.2019.TREATIES-XXVII.7.d of 4 November 2019. On 20 January 2021, the Government of the United States of America deposited its instrument of acceptance of the Agreement. See C.N.10.2021.TREATIES-XXVII.7.d of 20 January 2021.

The manufacturer's authorised representative in the EU is Springer Nature Customer Service Centre GmbH, Europaplatz 3, 69115 Heidelberg, Germany. If you have any concerns regarding our products, please contact ProductSafety@springernature.com

Printed and bound by CPI Group (UK) Ltd, Croydon, CR0 4YY
21/09/2026
02219962-0009